AF360914

Coral Reef Resilience

Coral Reef Resilience

Editors

Loke Ming Chou
Danwei Huang

MDPI • Basel • Beijing • Wuhan • Barcelona • Belgrade • Manchester • Tokyo • Cluj • Tianjin

Editors
Loke Ming Chou
National University of Singapore
Singapore

Danwei Huang
National University of Singapore
Singapore

Editorial Office
MDPI
St. Alban-Anlage 66
4052 Basel, Switzerland

This is a reprint of articles from the Special Issue published online in the open access journal *Journal of Marine Science and Engineering* (ISSN 2077-1312) (available at: https://www.mdpi.com/journal/jmse/special_issues/Coral_Reef_Resilience).

For citation purposes, cite each article independently as indicated on the article page online and as indicated below:

LastName, A.A.; LastName, B.B.; LastName, C.C. Article Title. *Journal Name* **Year**, *Volume Number*, Page Range.

ISBN 978-3-0365-0454-4 (Hbk)
ISBN 978-3-0365-0455-1 (PDF)

Contents

About the Editors

Loke Ming Chou obtained his Ph.D. in Zoology from the University of Singapore in 1976 and is currently Emeritus Professor at the National University of Singapore. He is actively engaged in research on coral reef ecology and integrated coastal management and has recently focused on reef restoration, particularly in highly turbid conditions of urban coasts. He is Honorary Fellow of the Singapore Institute of Biology and Fellow of the Singapore National Academy of Science.

Danwei Huang obtained his Ph.D. in Marine Biology from the Scripps Institution of Oceanography in 2012 and is currently Assistant Professor at the National University of Singapore. He is a coral reef biologist whose research focuses on the ecological and evolutionary history of corals and reef-associated organisms. He is interested in how the past has driven the present and can help predict the future of coral reef biodiversity, distribution, and health.

Journal of
Marine Science and Engineering

Editorial

Towards Coral Reef Resilience

Loke-Ming Chou [1,2,*] and Danwei Huang [1,2]

[1] Tropical Marine Science Institute, National University of Singapore, 18 Kent Ridge Road, Singapore 119227, Singapore; huangdanwei@nus.edu.sg
[2] Department of Biological Science, National University of Singapore, 14 Science Drive 4, Singapore 117543, Singapore
* Correspondence: tmsclm@nus.edu.sg

Received: 21 July 2020; Accepted: 22 July 2020; Published: 27 July 2020

Coral reef habitats provide valuable ecosystem services which have benefitted human society for millennia, but intense anthropogenic pressure, especially in the latter part of the last century, has resulted in widespread habitat degradation and loss of ecosystem services with severe environmental and societal consequences. Climate change impacts are expected to increase habitat stress further and compromise recovery and functioning of large swathes of reefs globally. Future scenarios range from almost total devastation to continued existence but in modified ecological states. Are coral reefs sufficiently resilient to withstand the changed environmental conditions of the future? This is an interesting aspect to consider. Numerous types of management responses have been attempted, and broadly include protection and restoration. Research is necessary to gain a better understanding of how reefs will respond to improved management as well as a changing climate. This would encompass various approaches to characterize and analyze reef responses from the molecular to community and habitat levels. Analyses may rely on spatially extensive and/or long-term monitoring data to detect and understand specific trends that are relevant to the formulation of novel management policy.

Research activity on reef resilience is increasing and will continue expanding as each piece of research adds some facet of information that helps to build the big picture. The five contributions (two reviews and three long-term monitoring assessments) in this volume provide interesting and critical information in this area.

New approaches to reef restoration built upon the reef gardening concept are proposed by Rinkevich [1]. They include improved coral gardening techniques, ecological engineering, assisted migration/colonization, assisted genetics/evolution, assisted microbiome, coral epigenetics, and coral chimerism. These components combined in an active reef restoration toolbox will reinforce the reef gardening restoration approach by helping to enhance coral resilience and adaptation to changing conditions and perhaps enhance our endeavor to secure a future for coral reefs.

From their review of the relationship between thermal stress and resilience, Carballo-Bolaños et al. [2] contend that natural response mechanisms of corals may not be sufficient to contribute to the habitat's ecological functioning if current greenhouse gas emission levels are not reduced. Corals mitigate thermal stress through various mechanisms including acclimatization, adaptation, and association with thermally tolerant endosymbionts. They resist thermal stress through molecular protective mechanisms such as heat shock proteins and antioxidant enzymes. Furthermore, each species of coral host and endosymbiont responds differently to thermal stress, highlighting the physiological diversity and complexity of the symbiotic partners. They conclude that thermal stress tolerance can be enhanced by approaches mentioned by Rinkevich [1], which include assisted migration/colonization, assisted evolution, ecological engineering, assisted genetics, and coral epigenetics.

The community analysis of Malauka'a fringing reef at Kāne'ohe Bay, Hawai'i, by Barnhill and Bahr [3] showed that corals acclimatized to a climate change-induced 0.96 °C increase over an 18-year period (2000 to 2018)—covering two major bleaching events—by retaining live coral cover and maintaining the two dominant species *Porites compressa* and *Montipora capitata*. However, a coral species compositional shift was attributed to the local loss of two species (*Pocillopora meandrina* and *Porites lobata*), replaced by a previously unrecorded species (*Leptastrea purpurea*). A significant decrease of the alga *Dictyosphaeria*, dominant in 2000, was seen together with the loss of *Gracilaria salicornia* and *Kappaphycus alvarezii*, accompanying an increase in non-coral substrate cover. The authors caution that while the reef system displayed resilience, the response may not be sufficiently swift to tolerate future temperature elevation and increasing bleaching frequency [3].

Keshavmurthy et al [4]. analyzed spatial and temporal (1986 to 2019) dynamics of corals in Kenting National Park (KNP), southern Taiwan, which features a fluctuating thermal environment induced by a branch of the Kuroshio Current and tide-induced upwelling that favored thermally-resistant corals, especially those close to the thermal effluent of a nuclear power plant. Major typhoons and bleaching caused coral cover fluctuations and spatial heterogeneity in coral cover recovery suggesting variable degrees of reef resilience between localities. Corals exposed to progressively warmer and fluctuating thermal environments possessed the ability to modify their endosymbiont community with a dominance shift to the thermally-tolerant *Durusdinium* spp. and reduce bleaching. Their study indicated that within a small geographical range with unique environmental settings and ecological characteristics, corals may be resilient to bleaching. They highlight the relevance of conservation efforts that are resilience-based to address climate change challenges [4].

An assessment of the resilience potential of inshore and offshore reef communities in the western part of the Gulf of Thailand by Sutthacheep et al. [5] over the last two decades showed that some sites in both areas had low resilience to bleaching. These reefs were also exposed to anthropogenic disturbances. However, some sites both inshore and offshore had high resilience potential based on bleaching survival rates although juvenile coral density was low. At most sites, juvenile coral density was not dependent on adult coral cover, particularly for *Acropora*. The authors recommend that resilience-based management should take into consideration natural processes that promote the resistance and recovery of corals, appropriate restoration efforts, and physical interventions such as shading during bleaching events.

Whether a reef is resilient to disturbance is challenging to uncover as it requires understanding of corals' susceptibility to and recovery from various stressors, which are often interacting with immense complexity. As presented in a number of the contributions here, evaluating the resilience of natural reefs requires long-term community data (>2 decades) and high-resolution environmental measurements. Due to the multiple factors involved, it is of no surprise that reef conditions and recovery outcomes post-disturbance are variable over relatively small spatial scales. These studies, and others emerging over the last decade, provide insight into the trajectories of coral reefs amidst more severe and frequent climate-related perturbations, including the possible scenario in which corals continue to survive and even dominate certain reefs, but with dramatic transformations at community to molecular levels. To anticipate these changes, restoration and management approaches must consider building resilience factors into coral reefs to future-proof these diverse and beneficial ecosystems.

Author Contributions: Both authors wrote this Editorial based on a synthesis of the findings presented in the five contributions to this special volume. All authors have read and agreed to the published version of the manuscript.

Funding: This research received no external funding.

Acknowledgments: The authors thank the contributors of the five papers for their positive and timely response to this special volume. The papers presented interesting observations especially from long-term monitoring and valuable insights on reef resilience, making our work of writing the editorial so much easier.

Conflicts of Interest: The authors declare no conflict of interest.

References

1. Rinkevich, B. The Active Reef Restoration Toolbox is a Vehicle for Coral Resilience and Adaptation in a Changing World. *J. Mar. Sci. Eng.* **2019**, *7*, 201. [CrossRef]
2. Carballo-Bolaños, R.; Soto, D.; Chen, C.A. Thermal Stress and Resilience of Corals in a Climate-Changing World. *J. Mar. Sci. Eng.* **2020**, *8*, 15. [CrossRef]
3. Barnhill, K.A.; Bahr, K.D. Coral Resilience at Malaukaa Fringing Reef, Kāne'ohe Bay, O'ahu after 18 years. *J. Mar. Sci. Eng.* **2019**, *7*, 311. [CrossRef]
4. Keshavmurthy, S.; Kuo, C.-Y.; Huang, Y.-Y.; Carballo-Bolaños, R.; Meng, P.-J.; Wang, J.-T.; Chen, C.A. Coral Reef Resilience in Taiwan: Lessons from Long-Term Ecological Research on the Coral Reefs of Kenting National Park (Taiwan). *J. Mar. Sci. Eng.* **2019**, *7*, 388. [CrossRef]
5. Sutthacheep, M.; Chamchoy, C.; Pengsakun, S.; Klinthong, W.; Yeemin, T. Assessing the Resilience Potential of Inshore and Offshore Coral Communities in the Western Gulf of Thailand. *J. Mar. Sci. Eng.* **2019**, *7*, 408. [CrossRef]

Journal of **Marine Science and Engineering**

Concept Paper

The Active Reef Restoration Toolbox is a Vehicle for Coral Resilience and Adaptation in a Changing World

Baruch Rinkevich

Israel Oceanography & Limnological Research, National institute of Oceanography, Tel Shikmona, P.O. Box 8030, Haifa 31080, Israel; buki@ocean.org.il; Tel.: +972-4-856-5275; Fax: +972-4-851-1911

Received: 23 May 2019; Accepted: 25 June 2019; Published: 28 June 2019

Abstract: The accelerating marks of climate change on coral-reef ecosystems, combined with the recognition that traditional management measures are not efficient enough to cope with climate change tempo and human footprints, have raised a need for new approaches to reef restoration. The most widely used approach is the "coral gardening" tenet; an active reef restoration tactic based on principles, concepts, and theories used in silviculture. During the relatively short period since its inception, the gardening approach has been tested globally in a wide range of reef sites, and on about 100 coral species, utilizing hundreds of thousands of nursery-raised coral colonies. While still lacking credibility for simulating restoration scenarios under forecasted climate change impacts, and with a limited adaptation toolkit used in the gardening approach, it is still deficient. Therefore, novel restoration avenues have recently been suggested and devised, and some have already been tested, primarily in the laboratory. Here, I describe seven classes of such novel avenues and tools, which include the improved gardening methodologies, ecological engineering approaches, assisted migration/colonization, assisted genetics/evolution, assisted microbiome, coral epigenetics, and coral chimerism. These are further classified into three operation levels, each dependent on the success of the former level. Altogether, the seven approaches and the three operation levels represent a unified active reef restoration toolbox, under the umbrella of the gardening tenet, focusing on the enhancement of coral resilience and adaptation in a changing world.

Keywords: climate change; reef restoration; gardening; ecological engineering; assisted migration/colonization; assisted genetics/evolution; assisted microbiome; epigenetics; chimerism

---

1. Introduction

Decades of continuous and substantial global climate change impacts, together with accumulated anthropogenic footprints on coral reefs, have demonstrated that, excluding a few remote reef sites, all major reefs suffer from accrued degradation, and a complete reshuffling of their biological diversity as they transform into less diverse ecosystems [1–3]. The abundance of corals and reef dwelling organisms has been impacted by escalating pressures and is continuously diminishing, while goods and services are failing [3] and biodiversity diminishes at ever growing rates, which are currently at 0.5–2% per year [4,5]. Climate change drives ocean warming and acidification, impacts overall physiological traits, triggers large-scale coral bleaching events, fuels tropical storms [6], slows reef calcification and growth, and impairs natural recruitment [7]. Moreover, devastating impacts are rapidly increasing in scale and intensity, bringing coral reefs to heightened eroded states globally, and affecting a decline in their ecological resilience capacities and adaptation to changing climate conditions. Globally, coral reef communities will most likely be in a state of flux for years to come (as many are already in), driven by different climate change drivers [8] with multiple stressors that act in tandem [9] and increase the risk of phase shifts into algal dominated reefs. Only a few reef sites exhibit some resistance to global climate change drivers [10].

As in other marine and terrestrial ecosystems, the rates of impact of climate change on species and populations are accelerating worldwide, calling for new forms of intervention. Furthermore, with the recognition that traditional measures (such as the creation of MPAs, reducing specific anthropogenic impacts, etc.) are not sufficient to cope with the combination of climate change/human footprints [4,11–13] the gloomy status of global reef ecosystems ignited the need for novel approaches that may accurately offset and mitigate the destructive impacts of global climate change, with alternative effective reef management and reef rehabilitation approaches. The initial idea was that these new approaches would be used to complement conservation efforts, allowing current reefs to provide ecosystem services under a range of future environmental conditions.

Probably the most effective among the emerging ideas, and the most widely used method, is the "gardening" approach for active reef restoration. This approach is based on 'principles', concepts and theories used in silviculture [13–19]. Taking into consideration coral reefs' inability to naturally recuperate without human intervention, the "gardening" concept, a fully employed active reef restoration, is a two-step process (the nursery phase dedicated to the development of large stocks of coral colonies in mid-water floating nurseries, followed by the transplantation phase where nursery-farmed coral colonies, which have reached suitable sizes, are out-planted onto degraded reef areas). The active "gardening" concept has emerged as an effective method [20], replacing the former less successful restoration approaches that focused on transplantation of coral colonies from a donor site onto a damaged site [13,21].

The terms 'active' and 'passive' restoration originated from forestation practices, which reflect two disparate broad categories [22]. 'Active' restoration is where human surrogate activities and practices directly help ecosystems recuperate or improve their state, while 'passive' restoration is when no human intervention is taken upon the reefs themselves, instead it focuses on reducing/eliminating anthropogenic impacts, allowing natural recuperation to lead the way to recovery [22,23]. One of the major benefits of active restoration is its critical role in reversing trajectories in ecosystems that are caught in dilapidated states [20,24]. Following this underlying principle, all key successful approaches for reef restoration (Table 1) use the 'active restoration' tactic, some of which harness natural processes such as assisted migration, epigenetics and coral chimerism (Table 1).

Table 1. The seven major research avenues added to the gardening approach for the creation of a climate adaptation toolkit (chosen references from the literature).

Avenue	Types of Coral Adaptation	Citations
Improved gardening methods	Development of various nursery types, adapted for a wide range of needs, improving coral self-attachment; using coral fragments without polyps; clustering of transplants improves outcomes; choosing favorable/improved substrates/coating, caging for recently settled spat—to enhance early post-settlement survival; spat feeding in *ex situ* nurseries for enhanced growth/survival; improved nursery maintenance by using environmentally friendly antifouling; increasing stocks of larvae from brooding coral species; improving seeding approaches; techniques for improved survival of coral propagules.	[25–44]
Ecological engineering	Use of herbivorous fish/invertebrates for improved nursery maintenance; animal-assisted cleaning; engineering of larval supply through transplantation of nursery-farmed gravid colonies; transplantation of ecological engineering species; development of larval hubs and 'artificial spawning hotspots'; tiling the reef; nubbin fusions for enlarged colonies; micro-fragmentation; serially positioning nurseries to create new mid-water coral biological corridors through stepping stone mechanisms; using dietary habits of grazers as biological controls of fouling macroalgae; large scale restoration acts; enhanced calcification/survival rates via seawater electrolysis.	[1,25–28,32,39,45–61]
Assisted migration/colonization	Moving species outside their historic ranges may mitigate loss of biodiversity in the face of global climate change.	[62–64]

Table 1. *Cont.*

Avenue	Types of Coral Adaptation	Citations
Assisted genetics/evolution	Enhanced coral adaptation, manipulating of algal symbionts to increase coral resistance to bleaching; using temperature tolerant genotypes; applying interspecific and intraspecific hybridization; using coral nurseries as genetic repositories.	[57,64–71]
Assisted microbiome	Adaptation by changing bacterial communities living in tissues, mucus layers and substrates to settle at the shortest timeframe of days/weeks; coral "microbial-therapy" and microbiome inoculation; improved nutrient cycles; contributing to coral host tolerance of thermal stress.	[72–75]
Epigenetics	Creation of novel alleles and traits that can better withstand environmental changes; developing resistance towards adverse conditions.	[46,76–83]
Chimerism	Enhanced growth and survival of spat/small colonies; countering the erosion of genetic and phenotypic diversity; high flexibility of chimeric entities on somatic constituents following changes in environmental conditions; the chimera synergistically presents the best-fitting combination of genetic components to environmental challenges; facilitating the healing of exposed coral skeletons	[84–91]

Since the short period that has elapsed since its inception, the employment of the gardening approach in a wide range of reef sites worldwide, has by now earned its credentials for (a) farming coral colonies from a large number of coral species (~ca 100) in mid-water nurseries, including massive, branching and encrusting forms; (b) establishing unlimited stocks of coral colonies in underwater nurseries; (c) the successful transplantation off nursery farmed coral colonies onto denuded reef areas, and (d) ensuring the low cost of farming and transplanting coral colonies [1,17]. However, this approach still lacks credibility in simulating restoration scenarios and trajectories that target specific goals. As such, additional restoration approaches were suggested and some have already been tested (Table 1), altogether creating a novel active reef restoration toolbox. Here, I'll summarize some of the major aspects and the hierarchy of these reef restoration avenues and approaches, which form the first toolbox to be used for enhancing coral resilience and coral adaptation in a changing world.

2. Defining the Toolbox

While active reef restoration techniques and their underlying fundamental principles are still under development, this discipline is challenged by the realization that reefs are already in transition, driven by differential species responses to environmental change, and that corals in the 'reef of tomorrow' should adapt to altering environmental conditions. The above infers that current basic methods for reef restoration are still insufficient to secure a future for coral reefs. This has prompted a surge in active restoration initiatives that can be divided into seven major research avenues added to the gardening approach (Table 1); each avenue is formulated in such a way as to guide an effective reef restoration tactic. Together they form a new reef restoration toolkit.

2.1. Improved Gardening Methodologies

As coral transplants show improved survival the larger they become, the early notion guiding the gardening approach was to develop coral colonies to a size that will significantly reduce mortality at transplantation sites. The midwater floating nurseries allow reduced competition for resources (substrate, light), better protection against predation pressures, provide improved conditions for reduced sedimentation and continuously increased water flow conditions for improved nutrition [26–28]. The working rationale has favored the demand for low-cost, low-tech reef restoration methodologies, with simple technical requirements that could be ubiquitously implemented anywhere worldwide [13–16,21]. This however is not sufficiently satisfactory, and the basic techniques that have been developed to maximize coral survival and productivity were supplemented by additional methodologies and technical approaches, all bundled under the title of 'improved gardening methodologies' (Table 1).

The literature in Table 1 reveals examples from a wide ranging, and continuously increasing, list of technological advancements, on almost every aspect of the coral gardening approach. This includes the development of various nursery types, adapted for a wide range of needs (such as the regular 'bed' nursery, the rope nursery, depth-adjustable nursery, nursery housing stock of large colonies, the larval dispersion hub nursery, and more (Figure 1) [1,26–28,45,46]; enhanced efficiencies for nursery maintenance, sustainability and yields (such as improved maintenance, harnessing herbivory by fishes and invertebrates as a parameter for positive maintenance feedbacks; spat feeding in ex situ nurseries for enhanced growth/survival; improved nursery maintenance by using environmentally friendly antifouling; caging for recently settled spat—to enhance early post-settlement survival; the use of coral fragments that lack polyps; the increasing stocks of larvae from brooding coral species; techniques for the improved survival of coral propagules), and more. The same goes for the transplantation phase, that has been augmented with improved methodologies, such as the development of different attachment procedures, improving coral self-attachment to substrates, clustering transplants for improved growth/survival outcomes, choosing favorable/improved substrates and coating materials, improved seeding approaches for enhanced settlement and early post-settlement survival, new seeding methodologies, augmenting post-transplantation growth and survival of juveniles via nutritional enhancement, maintaining/enhancing genotypic diversity, and more. While not yet tested for direct resilience and adaptation, the accumulated results suggest that improved gardening protocols not only enhance growth and survival at the nursery stage, but may have additional impacts on growth, survival and reproduction for years post-transplantation (e.g., [39,46,47]).

Figure 1. Three types of midwater floating nurseries, the first step of the "gardening" tenet. Nurseries are adapted for various transplantation needs and practices. (**a**,**b**), the regular 'bed' nursery, where corals (usually mono-species cultures) are directly farmed on the nursery base. (**a**) a short period after inception, where most of the mesh-base of the nursery is still seen (*Acropora formosa*, Bolinao, the Philippines). (**b**) a 'bed' nursery completely covered with *Montipora digitata* colonies (Bolinao, the Philippines). (**c**) a classical floating nursery. The nursery substrate is made of a rope net (sized 10 × 10 m). Coral nubbins

are glued onto plastic pins (9 cm long, 0.3–0.6 cm wide leg, and 2 cm diameter "head") and are inserted into plastic nets stretched over PVC frames (30 × 50 cm). Frames with corals are tied to the nursery substrate (Eilat, Israel). This type of nursery allows for a pre-planned transplantation protocol, where each coral colony has its own 'pot' (the plastic pin) and the transplantation protocol considers the attached pin, with limited stress to the growing coral. An established nursery attracts fish and reef associated invertebrates recruited from the plankton. (**d**) Rope nursery (Bolinao, the Philippines). This nursery accommodates small coral fragments inserted within the rope threads, creating an easily constructed nursery bed that is transplanted together with the developing corals. Photos: a,b,d = G. Levy, c = S. Shafir.

2.2. Ecological Engineering

Ecological engineering is defined as: "the design of sustainable ecosystems that integrate human society with its natural environment for the benefit of both" [92]. It involves not only the restoration of ecosystems that have been noticeably altered by either anthropogenic impacts and/or global climate change drivers, but also reflects the emerging scientific discipline that is associated with the development of sustainable new and/or hybrid ecosystems, which have human and ecological significance, providing (when possible) equivalent levels of goods and services as the original ecosystems.

As noted earlier [17] the active gardening approach can be regarded as a ubiquitous ecological engineering platform for reef restoration measures performed on a global scale, having properties that incorporate ecological engineering aspects and tools under a common scientific umbrella (e.g., [39,46,47,84]), including the use of species (corals, fish, other invertebrates) that are allogenic and autogenic ecosystem engineers. This is of specific importance since climate change drivers may hinder the ecological engineering capacities of scleractinian corals as primary reef ecosystem engineers [93]. Clearly, this requires a comprehensive understanding of the engineering capabilities that may be associated with reef restoration approaches, and of the ways ecological engineering species function as reef ecosystem engineers.

Both scientific notions, 'ecological engineering' and 'ecosystem restoration', while representing distinct disciplines [94], are widely used together in terrestrial environments to repair a number of deterioration scenarios [92,94,95]. While 'ecological engineering' provides more predictable outcomes with higher functionalities associated with the chosen ecosystem services, 'ecological restoration' tends to produce higher diversity outcomes, which are aimed at long-term recovery of lost ecosystem services. Principles of both disciplines are primarily intermingled in large scale restoration efforts [94]. Focusing on coral reef ecosystems, ecological engineering tactics, together with restoration of degraded reef habitats, are increasingly recognized as valuable tools, primarily in association with the gardening approach [17,21,39,46,47,84]. It has been also suggested [47] that integrating functional considerations into transplantation acts, such as in the use of allogenic and autogenic engineer species, could improve the impacts of restoration on reef biodiversity.

The literature in Table 1 offers examples from the wide-ranging and increasing number of ecological engineering approaches, covering various aspects of the coral gardening tenet. The prevailing belief predicts that herbivory by fishes and invertebrates (primarily sea urchins and gastropods) is the cornerstone of the developed complex ecological networks that suppress macroalgal cover, minimize coral–algal competition, increasing coral growth and recruitment and dictating coral-dominated reefs' health levels. As a result, much attention has been devoted to the use of herbivorous organisms for improved nursery maintenance, for animal-assisted cleaning and for adapting dietary habits of grazers as biological controls of fouling macroalgae in coral nurseries [25–27,61]. As a matter of fact, in the Eilat (Red Sea) nursery, herbivores like the fish *Siganus rivulatus* and the sea urchin *Diadema setosum* controlled algal growth by virtue of intensive grazing [25]. This becomes even more relevant with the forecasted global climate change impacts on grazing kernels (e.g., [96]). In the same way, coralivorous species in the Eilat nursery [28] could be effectively eliminated by a top down control reliant on fish predation (mainly *Thalassoma rueppellii* and *T. lunare*).

The recently developed ecological engineering approaches are also engaged in various reproductive activities and planula larvae aspects. Examples are the engineering of larval supply through transplantation of nursery-farmed gravid colonies [46], the establishment of coral nurseries as larval dispersion hubs and as 'artificial spawning hotspots' [1,17,44,47,97], and the enhancement of larval survival/growth under nursery conditions [32,33,58]. Several entire-reef ecological engineering aspects involved are for example: the selection of coral species for reef restoration while considering their autogenic/allogenic engineering properties [39], serially positioning nurseries to create novel mid-water biological corridors for larval recruitment through stepping stone mechanisms [17], enhancing calcification and survival rates through electrolysis in seawater [48–50], micro-fragmentation of coral colonies for various purposes such as tiling the reefs, and the creation of large colonies within short time periods [53,59,60] versus nubbins/spat fusions for enlarged colonies [53,84], and more. All the above mentioned may enhance efficiency rates of the gardening restoration approach in combating the impacts of global climate change [98].

2.3. Assisted Migration/Colonization

Climate change is causing spatial-temporal shifts in environmental conditions, challenging species that are unable to relocate to suitable environments, thus increasing their risk of extinction. Human directed (Table 1) and natural movements of coral species outside their historic ranges ('assisted migration/colonization' and 'natural range expansion', respectively) into more favorable sites, may mitigate the loss of biodiversity in the face of global climate change [62]. Indeed, natural poleward range expansion of corals has been widely documented, from recent fossil records where *Acropora*-dominated reefs extended along the Florida coast as far north as Palm Beach County [99] and from Australian Pleistocene reefs [100], to the last 80 years of national records from Japanese temperate areas, where key reef formation species revealed speeding poleward range expansions of up to 14 km/year [101,102] and to coral species range extensions in the Eastern and Western Australian coasts [103,104]. While these and other studies support the notion that gradual warming seems to drive range extensions of tropical reef fauna into temperate areas, other studies [105] noted that the dose of photosynthetically available radiation over winter can severely constrain such latitudinal coral habitat expansions.

As for assisted migration/colonization, this conservation strategy has been considered not only for the relocation of species, populations, genotypes, and/or phenotypes to sites beyond their historical distribution, but also for species whose ranges have become highly fragmented [62]. While some studies suggest that assisted colonization is viable due to the introduction of novel, and/or relaxed selection, such operations may lead to an unintended evolutionary divergence [106], which is known to generally yield a low success rate [107] and which is further less effective for species that rely on photoperiodic and thermal cues for development [108]. All the above mentioned is associated with reduced ecosystem services and diminished ecological complexity as characteristics of this approach [17]. An additional criticism raised is that the employment of assisted colonization with rare or endangered species (like the Caribbean *Acropora* species; also, the introduction of pathogens and predators to new locations) poses a great risk for them as well as for the recipient locations [109].

Harnessing the natural phenomenon of coral colonies that raft on floating objects for thousands of kilometers [110], and the natural range expansion of coral species, human intervention through assisted colonization is considered a part of the toolkit of active reef restoration [1,17]. Claims have been made [63,64] that Arabian/Persian Gulf corals, which are already surviving in thermal conditions forecasted to prevail in the future in most tropical reefs, can be considered as a source for assisted migration to the tropical Indo-Pacific. Inter-population hybridizations of gravid colonies adapted to cooler versus warmer temperature areas (such as in the case of *Acropora millepora* from the Great Barrier

Reef, Australia [111]) may also be a promising candidate for the assisted migration management of offspring.

2.4. Assisted Genetics/Evolution

Assisted evolution/genetics has recently been defined as: "a conservation strategy that involves manipulating the genes of organisms in order to enhance their resilience to climate change and other human impacts" [112]. Assisted evolution/genetics has come to the forefront because climate change has been shown to outpace natural rates of evolution. This may span a wide range of aspects that target either the coral colonies and/or their algal symbionts, including: enhanced coral adaptation; manipulation of algal symbionts to increase coral resistance to bleaching; use of temperature tolerant genotypes to mitigate new environmental challenges; applying interspecific and intraspecific hybridization efforts; using coral nurseries as genetic repositories; and more (Table 1). With regards to the topic of this manuscript, gaining a better understanding of adaptation at the genetic level would clearly benefit coral restoration projects [113,114]. Over the short and intermediate terms, corals may adapt to changing environmental conditions by transforming holobiont (coral-algal) properties [65] whereby algal symbiont communities are changed into types/species/clades that enhance the stress tolerance of the host coral. In the long term, changes may occur within the genetic blueprint of the coral colonies, through supportive breeding plans within populations, outcrossing between populations and hybridization between closely related species.

Resulting from the exceptional genetic variability that naturally exists within the endosymbiotic dinoflagellate algae of the family Symbiodiniaceae, much of the assisted evolution/genetics work has been concentrated on manipulating algal species residing within tissues of coral colonies from the same species. This is based on the rationale that seeding less resilient corals with temperature adapted algal variants would provide a management/restoration tool to reduce bleaching and mortality of corals subjected to temperature stress [67,69,71,113,115]. However, it must be emphasized that while the literature attests that corals may naturally experience changes in symbiont communities following bleaching episodes, directed manipulations of adult corals in favor of more thermos-tolerant symbionts have only been achieved in the laboratory to date [116].

Following the observation that naturally resilient corals are scarce, genetic manipulation of coral communities under stress conditions is suggested more and more. This includes moving more resilient coral colonies to vulnerable areas within and outside of their species distribution areas, associated with the assisted migration/colonization tenet [63,64,111,112]. Another approach is the adoption of breeding programs within populations, outcrossing between populations and hybridizing closely related species [70]. The current research, however, is still at the proof-of-concept stage. While natural hybridization is known in some scleractinian corals, such as the genus *Acropora*, the applicability of this approach, the fitness of offspring from such outcrossing/hybridization programs in the field, as well as the establishment of successful F2 progenies and their reproductive activities, are all yet to be investigated.

Another assisted genetics/evolution approach is based on the understanding and evidence [81] that coral populations in current reefs embrace a reservoir of alleles preadapted to a wide range of future challenges, such as higher temperatures. This outcome is still poorly documented in measurable parameters and effects. However, the findings point to the potentiality for a rapid evolutionary response to climate change, and the legitimate inclusion of this phenomenon as an efficient restoration tool. This is also connected to the suggestion of using coral nurseries as repositories for genetic material that would have otherwise been lost from reef sites, preserving genotypes for future restoration efforts [66]. All the above mentioned is in addition to the consideration of coral nurseries as applied tools to capture and harvest coral larvae, to increase genetic diversity or to grow mature breeding corals for larval production and the seeding of degraded reefs [1,17,32,33,44,47,58,97].

2.5. Assisted Microbiome

The assisted microbiome tenet, aligned with the assisted genomics/evolution view, is led by the coral probiotic hypothesis [72] for enhancing the adaptation potential of corals to changing environmental conditions through changes in associated bacterial communities. Using this tenet as adaption and restoration tools (Table 1), it has been suggested that microbiome manipulation may alter the coral phenotypes, and subsequently the entire colonies' fitness to withstand environmental challenges [73–75,117].

While at present little is known about the mechanisms related to the "probiotic" protection provided by the coral microbiome, and a key uncertainty exists about the feasibility of manipulating microbes to enhance coral tolerance [73], microbial symbionts were suggested as contributors to the physiology, development, health and immunity of corals, and as a tool to facilitate nutrient cycling and nutrition in general [116,117]. Following this rationale, the manipulation of microbiome communities has been suggested as a key strategy to 'engineer' coral phenotypes. However, the ecosystem functioning of bacteria inoculation necessitates further work, as targeted actions are problematic to design without the needed baseline studies [116].

2.6. Epigenetics

Organism responses to any environmental challenge develop through either genetic change (e.g., allele frequency alternations between generations, mutational accumulation) and/or nongenetic (i.e., epigenetics) processes. Epigenetics refers to external modifications in genes (e.g., methylation, acetylation, histone modifications and small RNAs; without any modification in gene sequences) that cause change in gene expression. The literature attests that many of the environmentally induced epigenetic changes are, as a matter of fact, heritable [118], thus facilitating the acceleration of adaptation processes.

It is generally assumed that epigenetics allows corals a greater ability to buffer the impacts of environmental changes and of various stress conditions (Table 1), by fine-tuning gene expression, thereby providing additional time for genetic adaptation to occur. A recent study [83] has revealed that epigenetics significantly reduced spurious transcription in the Indo-Pacific coral *Stylophora pistillata*, diminishing transcriptional noise by fine-tuning gene expressions and causing widespread changes in pathways regulating cell cycle and body size, with impacts on cell and polyp sizes as well as skeletal porosity. In a similar way, probable epigenetic signatures (a) imposed diminished bleaching responses when comparing two of the most severe episodes (17 y period) of global-scale seawater temperature anomalies [79], and (b) assisted transplanted gravid coral colonies to release an order of magnitude more coral larvae than local colonies for at least 8 reproductive seasons post transplantation ([46]; unpubl.). Coral epigenetics as a management tool, alleviating impacts of global climate change on reef corals, and as a potential tool for improving reef restoration outcomes, has further gained support from studies showing links between coral adaptation and epigenetics [46,77–83].

Interestingly however, epigenetic changes may also be induced under 'healthy', more pampered situations, such as under parental care and improved nutrition [119–121]. Various epigenetic impacts have already been suggested to develop in coral colonies or coral fragments subject to different environmental conditions [46,77,83,84], most interesting of all are the impacts on heightened long-term coral reproductive capabilities [46]. Thus, favorable biological and physical conditions at the nursery stage, including: optimal light conditions, increased water flow, minimized sedimentation, enhanced planktonic supply, reduced intra- and interspecific competition, and controlled corallivory [15,26–28, 45,58,122], may impose lasting epigenetic changes on fitness and on ecological traits of transplanted corals, enhancing their ability to counter global climate change impacts and other less-favorable environmental conditions. It should be noted however that while meriting further experimental investigation, the discipline of epigenetics and epigenetic impacts in corals is still in its infancy.

2.7. Coral Chimerism

A new potential tool in reef restoration (Table 1) that stems from the phenomenon of coral chimerism (Figure 2 [85]). The coral chimera is a biological entity that simultaneously consists of cells originating from at least two sexually-born conspecifics, a natural tissue transplantation phenomenon intermingling complex ecological and evolutionary mechanisms and concepts [123,124]. With regards to reef restoration, coral chimerism is presented as one of the best applied tools for accelerating adaptive responses to global climate change impacts [85], thus improving reef restoration tactics. The adaptive qualities are based on the suggestion that coral chimerism counters the erosion of genetic and phenotypic diversity, by presenting high flexibility on somatic constituents following changes in environmental conditions. This enables all partners in a chimera to synergistically present the best-fitting combination of genetic components to the environment [85,123,124]. In most cases, chimerism in corals is restricted to specific short windows at early ontogenic stages [125,126] and chimeric impacts are evident from early stages of development [86].

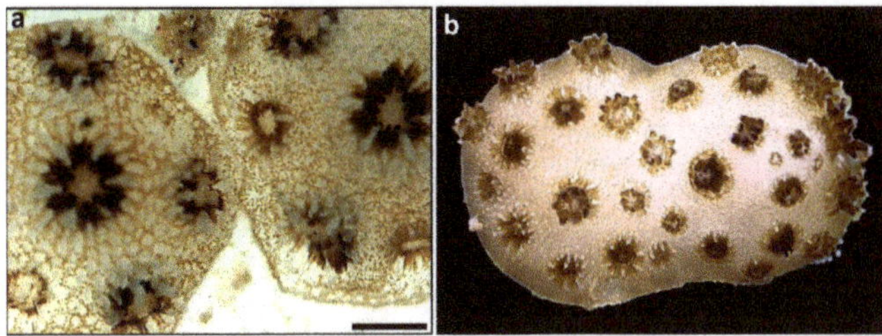

Figure 2. Coral chimerism. (**a**) Two contacting young spats (about 1 month old) of the Red Sea branching coral *Stylophora pistillata*, during the process of fusion (bar = 2 mm); (**b**) a several months old chimera of *Stylophora pistillata*, before the initiation of up-growing branches. Morphologically undistinguished area of fusion.

The literature documents a wide range of ecological advantages and benefits incurred to coral chimeras. Chimerism endows the chimeric entity, primarily at early life-history stages, with an instant survival advantage, like enhanced growth rates by virtue of the abrupt increase in size when the two organisms merge [84,86–88], and facilitation of the healing of exposed coral skeletons by enhanced preferential gregarious settlement of coral planulae [89]. The development of asexual chimeric coral planulae [90] together with the phenomenon of planulae fusion in the water column [88,91] may further mitigate the loss of genetic diversity of small colonizing populations [85,90].

The phenomenon of coral chimerism (Figure 2) is probably one of the least explored potential pathways corals take to buffer the impacts of capricious environmental conditions. Studying coral chimerism is not a trivial task and much has to be investigated before a better understanding can be achieved regarding this unique natural phenomenon and its inclusion in the coral restoration toolbox, another added facet to the gardening approach for active reef restoration [1,17].

3. Discussion

Ecological restoration is broadly defined as: 'the process of assisting the recovery of an ecosystem that has been degraded, damaged, or destroyed' [127], and is becoming the major ubiquitous strategy for increasing ecosystem services, as well as for reversing biodiversity decline. As a relatively new discipline it is fraught with hindrances, which is to be expected [128]. In contrast, the science of restoration ecology (primarily the facets that deal with terrestrial ecosystems), has rapidly developed over the past century, maturing into a cohesive body of theory that is backed by an established toolbox of restoration practices. Notwithstanding the growing interest in ecological restoration, the added challenges posed by climate change further reveal that the available adaptation toolkit associated with ecological restoration is still meager [129]. This is also emphasized in the coral restoration arena, a field that has not yet developed to the level of scientific maturity comparable to that of terrestrial ecological restoration [1,17].

On top of anthropogenic activities, climate change significantly challenges the concepts, practices and outcomes of ecological restoration. It is now more than a decade since the realization that it makes less sense to establish current restoration approaches on historical references, as they are all under the influence of rapidly changing climate regimes. Although historical references are of interest, they are less useful as ways to establish direct objectives [127]. Furthermore, the forecasted climate change scenarios will pose further challenges, some of which are yet to be experienced. Additionally, restoration efforts will have to address, in addition to restitution of biodiversity and ecosystem services, the ecosystem's resilience in the face of anticipated climate change scenarios [114,130].

This manuscript deals with the currently developing active reef restoration toolbox, used to enhance coral resilience and adaptation in a changing world. Seven classes of avenues and tools were described (Table 1) and discussed, including: the improved gardening methodologies, ecological engineering approaches, assisted migration/colonization, assisted genetics/evolution, assisted microbiome, coral epigenetics and coral chimerism. These tools are further classified into three levels of operation (Figure 3), each is based on the success of the former level, altogether compiling the most current active reef restoration toolbox. This toolbox is based on the rational and methodologies developed for the 'coral gardening' concept [13–19,21,26–28].

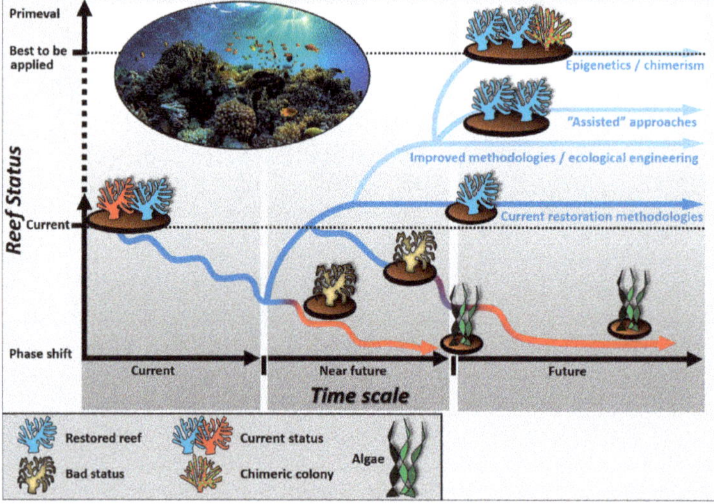

Figure 3. A theoretical illustration depicting how the seven classes of the suggested novel avenues and tools (improved gardening methodologies, ecological engineering approaches, assisted migration/colonization, assisted genetics/evolution, assisted microbiome, coral epigenetics, coral chimerism), further classified into three operation levels, compiling a unified active reef restoration toolbox, under the umbrella of the gardening tenet. Using the currently available restoration methodologies (based on the gardening approach) reef statuses that are anticipated to decline (the red trajectory towards the near future) are improving, or not ([the red trajectory towards the future] depending on the level of stress imposed by anthropogenic activities and climate change drivers). The next evolved level of progress in reef status is achieved by applying improved methodologies and ecological engineering approaches. They may maintain an improved reef status, but not the desirable advanced state. Yet, this level provides the ground for the operational level of 'assisted' approaches and the apex operational level of epigenetics and chimerism approaches, altogether maximizing reef statuses and enhancing coral resilience and adaptation in a changing world, developing to the 'best to be applied' status with current research avenues, yet not approaching the primeval reef status.

The basic and first level (Figure 3) includes two classes of tools, the improved gardening methodologies and the ecological engineering approaches, which are aimed at further enhancing the efficiency of the coral restoration approach, towards the development of sustainable ecosystems that have human and ecological significance. The research in both classes of coral restoration tools, either on the nursery or the transplantation phases, is highly active, performed in various reefs worldwide on a wide range of coral species, and various new approaches and methodologies are frequently suggested and tested. In addition to maximizing the survival and growth rates of corals in the nursery and after transplantation, the new approaches (primarily the ecological engineering approaches) tackle major issues in reef restoration. These include the phase-shifting of coral reef surfaces from turf algae back to coral dominated layers [60], the creation, within very short time periods, of large coral colonies of ecological importance [53,59,60], and the establishment of new biological corridors through stepping stone mechanisms [17] just to name a few of the ramifying approaches.

The second level (Figure 3) includes the three 'assisted' approaches (assisted migration/colonization, assisted genetics/evolution, and assisted microbiome). This level of operation represents restoration strategies and approaches that shift in theory and in practice from former approaches reliant on reference points and historically based goals, towards a common focus on "process-oriented configurations" [130]. The assisted approaches are still either at a conceptual level, or first laboratory trials, and are challenged by the need to guide the transition towards ecosystem states that can maintain key functions and values in a changing environment. For example, the assisted migration/colonization approach as developed may result in a new ecosystem with reduced services and diminished ecological complexity [17]. The assisted genetics/evolution approach is still at the proof-of-concept stage [116], while the assisted microbiome approach and the suggested activities therein, are still problematic to design as they lack the needed baseline studies [116]. The 'assisted' approaches hinge on successful active restoration methodologies, such as nursery grown colonies and transplantation tactics. It is most likely that much of the 'assisted' approaches will be shaped and intermingled in the future with other ecological engineering approaches to form a toolkit, aimed at achieving an improved ecologically-based restoration strategy. Thus, it is envisaged that neither one of the assisted approaches will stand by itself as an independent restoration strategy.

The third operational level (Figure 3) includes the two approaches of coral epigenetics and coral chimerism. While the success in either approach depends on the rationale and methodologies developed for the 'coral gardening' concept, and on the supplementary ecological engineering toolkit, each approach is based on a well-established biological phenomenon with considerable ecological and evolutionary perspectives. Employing the coral epigenetics tool may provide extra tolerance in case of subsequent re-exposure of the organism (or its progeny) to similar or even harsher conditions. At this stage, most studies on the subject were performed under laboratory conditions or on evaluations of coral responses from the field [77–79,81–83] but there is also documentation for novel phenotypic attributes developed following human manipulation under field conditions (increased growth rates of corals, long term enhancement of reproduction output [46]). Employing the coral chimerism tool may further provide cumulative levels of adaptation, as they are expressed by a naturally occurring phenomenon [84–91,125,126].

Coral chimerism (Figures 2 and 3) has already been discussed as a potential evolutionary rescue instrument, reliant on the premise that it may compensate for the immediate need for genetic change [85]. In a similar way, an epigenetic modification can facilitate evolutionary rescue through the creation of novel phenotypic variants [131]. Thus, both instruments may provide coral populations with the resilience to persist through periods of environmental change. Both instruments, alone or in combination, have the potential to facilitate faster adaptation rates and improved adaptation, than those exhibited in traditional genetic mutations, and thus merit special attention.

It should be noted, however, that risks involved in the application of some of the tools are not yet well defined and that the potential of unknown costs versus perceived benefits assigned to the tools should be evaluated [106–108,116]. These include costs for selective breeding that may lead to reduced genetic variability, and for increased sensitivity of coral populations to other climate change drivers, the introduction of pathogens and predators via coral transplantation [109], and for the flawed allocation of limited human, institutional and financial resources [17,116]. Another topic not addressed here is the scale of future restoration measures at the changing world. While the coral gardening-toolbox could serve as a ubiquitous ecological engineering platform for restoration on a global scale, it is yet facing the most imperative challenge to document restoration manipulations at regional/global levels [17], to determine that the gardening approach indeed supports sustainable coral reefs at large scales. Indeed, results already noted that large-scale coral restoration may have a positive influence on coral survivorship [132], recruitment rates and juvenile density [56]. These acts may further be aided by novel tools, like remote sensing technology [133].

Cumulatively, climate change and anthropogenic impacts pose major challenges for the development of effective tools, not only assessing levels of degradation in reef ecosystems under varying states of alteration, but also for the development of rationales and methodologies to efficiently restore degrading reefs. Based on principles, concepts and theories from silviculture, the "gardening" concept of active reef restoration [13–19,21,26–28] has not only laid the foundation for reef restoration, but is now developing through several seemingly separate approaches (improved gardening methodologies, ecological engineering approaches, assisted migration/colonization, assisted genetics/evolution, assisted microbiome, coral epigenetics and coral chimerism) that are divided here into three operational levels, altogether representing the unified active reef restoration toolbox under the umbrella of the gardening tenet to focus on the development of coral resilience and adaptation in a changing world. This may lead to new policies that will be integrated with other efforts to scale up reef restoration efforts into a global measure embedded within integrated governance structures.

Funding: This research was funded by the AID-MERC program (no M33-001), by the North American Friends of IOLR (NAF/IOLR), by the JNF and by the Israeli-French high council for scientific & technological research program (Maïmonide-Israel).

Acknowledgments: This study was supported by Thanks are due to G. Paz for drawing Figure 3.

Conflicts of Interest: The author declares no conflicts of interest.

References

1. Rinkevich, B. Climate change and active reef restoration—Ways of constructing the "reefs of tomorrow". *J. Mar. Sci. Eng.* **2015**, *3*, 111–127. [CrossRef]
2. Sully, S.; Burkepile, D.E.; Donovan, M.K.; Hodgson, G.; Van Woesik, R. A global analysis of coral bleaching over the past two decades. *Nat. Commun.* **2019**, *10*, 1264. [CrossRef] [PubMed]
3. Jones, K.R.; Klein, C.J.; Halpern, B.S.; Venter, O.; Grantham, H.; Kuempel, C.D.; Shumway, N.; Friedlander, A.M.; Possingham, H.P.; Watson, J.E. The location and protection status of Earth's diminishing marine wilderness. *Curr. Biol.* **2018**, *28*, 2506–2512. [CrossRef] [PubMed]
4. Bruno, J.F.; Selig, E.R. Regional decline of coral cover in the Indo-Pacific: Timing, extent, and subregional comparisons. *PLoS ONE* **2007**, *2*, e711. [CrossRef] [PubMed]
5. De'ath, G.; Fabricius, K.E.; Sweatman, H.; Puotinen, M. The 27-year decline of coral cover on the great barrier reef and its causes. *Proc. Natl. Acad. Sci. USA* **2012**, *109*, 17995–17999. [CrossRef] [PubMed]
6. Trenberth, K.E.; Cheng, L.; Jacobs, P.; Zhang, Y.; Fasullo, J. Hurricane Harvey links to ocean heat content and climate change adaptation. *Earths Future* **2018**, *6*, 730–744. [CrossRef]
7. Hughes, T.P.; Kerry, J.T.; Baird, A.H.; Connolly, S.R.; Chase, T.J.; Dietzel, A.; Hill, T.; Hoey, A.S.; Hoogenboom, M.O.; Jacobson, M.; et al. Global warming impairs stock—Recruitment dynamics of corals. *Nature* **2019**, *568*, 387–390. [CrossRef] [PubMed]
8. IPCC. Climate Change 2014: Synthesis Report. Contribution of Working Groups I, II and III to the Fifth Assessment Report of the IPCC. In *Proccedings of the Intergovernmental Panel on Climate Change*; IPCC: Geneva, Switzerland, 2014.

9. Fong, C.R.; Bittick, S.J.; Fong, P. Simultaneous synergist, antagonistic and additive interactions between multiple local stressors all degrade algal turf communities on coral reefs. *J. Ecol.* **2018**, *106*, 1390–1400. [CrossRef]

10. Fine, M.; Cinar, M.; Voolstra, C.R.; Safa, A.; Rinkevich, B.; Laffoley, D.; Hilmi, N.; Allemand, D. Coral reefs of the Red sea-challenges and potential solutions. *Reg. Stud. Mar. Sci.* **2019**, *25*, 100498. [CrossRef]

11. Rinkevich, B. Management of coral reefs: We have gone wrong when neglecting active reef restoration. *Mar. Pollut. Bull.* **2008**, *56*, 1821–1824. [CrossRef]

12. Miller, K.I.; Russ, G.R. Studies of no-take marine reserves: Methods for differentiating reserve and habitat effects. *Ocean Coast. Manag.* **2014**, *96*, 51–60. [CrossRef]

13. Rinkevich, B. Restoration strategies for coral reefs damaged by recreational activities: The use of sexual and asexual recruits. *Restor. Ecol.* **1995**, *3*, 241–251. [CrossRef]

14. Rinkevich, B. Steps towards the evaluation of coral reef restoration by using small branch fragments. *Mar. Biol.* **2000**, *136*, 807–812. [CrossRef]

15. Rinkevich, B. Conservation of coral reefs through active restoration measures: Recent approaches and last decade progress. *Environ. Sci. Technol.* **2005**, *39*, 4333–4342. [CrossRef] [PubMed]

16. Rinkevich, B. The coral gardening concept and the use of underwater nurseries; lesson learned from silvics and silviculture. In *Coral Reef Restoration Handbook*; Precht, W.F., Ed.; CRC Press: Boca Raton, FL, USA, 2006; pp. 291–301.

17. Rinkevich, B. Rebuilding coral reefs: Does active reef restoration lead to sustainable reefs? *Curr. Opin. Environ. Sustain.* **2014**, *7*, 28–36. [CrossRef]

18. Epstein, N.; Bak, R.P.M.; Rinkevich, B. Strategies for gardening denuded coral reef areas: The applicability of using different types of coral material for reef restoration. *Restor. Ecol.* **2001**, *9*, 432–442. [CrossRef]

19. Epstein, N.; Bak, R.P.M.; Rinkevich, B. Applying forest restoration principles to coral reef rehabilitation. *Aquat. Conserv.* **2003**, *13*, 387–395. [CrossRef]

20. Rinkevich, B. The quandary for active and passive reef restoration in a changing world. In *Active Coral Reef Restoration: Techniques for a Changing Planet*; Vaughan, D., Ed.; J. Ross Publishing: Fort Lauderdale, FL, USA, in press.

21. Lirman, D.; Schopmeyer, S. Ecological solutions to reef degradation: Optimizing coral reef restoration in the Caribbean and Western Atlantic. *Peer J.* **2016**, *4*, e2597. [CrossRef]

22. Bradshaw, A.D. Underlying principles of restoration. *Can. J. Fish. Aquat. Sci.* **1996**, *53* (Suppl. 1), 3–9. [CrossRef]

23. Holl, K.D.; Aide, T.M. When and where to actively restore ecosystems? *For. Ecol. Manag.* **2011**, *261*, 1558–1563. [CrossRef]

24. Clewell, A.F.; Aronson, J. Motivations for the restoration of ecosystems. *Conserv. Biol.* **2006**, *20*, 420–428. [CrossRef] [PubMed]

25. Shafir, S.; Van Rijn, J.; Rinkevich, B. Steps in the construction of underwater coral nursery, an essential component in reef restoration acts. *J. Mar. Biol.* **2006**, *149*, 679–687. [CrossRef]

26. Shafir, S.; van Rijn, J.; Rinkevich, B. A mid-water coral nursery. In Proceedings of the 10th International Coral Reef Symposium, Okinawa, Japan, 28 June–2 July 2004; pp. 1674–1679.

27. Shafir, S.; Rinkevich, B. The underwater silviculture approach for reef restoration: An emergent aquaculture theme. In *Aquaculture Research Trends*; Schwartz, S.H., Ed.; Nova Science Publications: New York, NY, USA, 2008; pp. 279–295.

28. Shafir, S.; Rinkevich, B. Integrated long term mid-water coral nurseries: A management instrument evolving into a floating ecosystem. *Univ. Maurit. Res. J.* **2010**, *16*, 365–379.

29. Shafir, S.; Abady, S.; Rinkevich, B. Improved sustainable maintenance for mid-water coral nursery by the application of an anti-fouling agent. *J. Exp. Mar. Biol. Ecol.* **2009**, *368*, 124–128. [CrossRef]

30. Baria, M.V.B.; Guest, J.R.; Edwards, A.J.; Aliño, P.M.; Heyward, A.J.; Gomez, E.D. Caging enhances post-settlement survival of juveniles of the scleractinian coral *Acropora tenuis*. *J. Exp. Mar. Biol. Ecol.* **2010**, *394*, 149–153. [CrossRef]

31. Liñán-Cabello, M.A.; Flores-Ramírez, L.A.; Laurel-Sandoval, M.A.; Mendoza, E.G.; Santiago, O.S.; Delgadillo-Nuño, M.A. Acclimation in *Pocillopora* spp. during a coral restoration program in Carrizales Bay, Colima, Mexico. *Mar. Freshw. Behav. Physiol.* **2011**, *44*, 61–72. [CrossRef]

32. Linden, B.; Rinkevich, B. Creating stocks of young colonies from brooding-coral larvae, amenable to active reef restoration. *J. Exp. Mar. Biol. Ecol.* **2011**, *398*, 40–46. [CrossRef]

33. Linden, B.; Vermeij, M.J.A.; Rinkevich, B. The coral settlement box: A simple device to produce coral stock from brooded coral larvae entirely in situ. *Ecol. Eng.* **2019**, *132*, 115–119. [CrossRef]

34. Nakamura, R.; Ando, W.; Yamamoto, H.; Kitano, M.; Sato, A.; Nakamura, M.; Kayanne, H.; Omori, M. Corals mass-cultured from eggs and transplanted as juveniles to their native, remote coral reef. *Mar. Ecol. Prog. Ser.* **2011**, *436*, 161–168. [CrossRef]

35. Nedimyer, K.; Gaines, K.; Roach, S. Coral Tree Nursery©: An innovative approach to growing corals in an ocean-based field nursery. *Aquac. Aquar. Conserv. Legis.* **2011**, *4*, 442–446.

36. Cooper, W.T.; Lirman, D.; VanGroningen, M.P.; Parkinson, J.E.; Herlan, J.; McManus, J.W. Assessing techniques to enhance early post-settlement survival of corals in situ for reef restoration. *Bull. Mar. Sci.* **2014**, *90*, 651–664. [CrossRef]

37. Tebben, J.; Guest, J.R.; Sin, T.M.; Steinberg, P.D.; Harder, T. Corals like it waxed: Paraffin-based antifouling technology enhances coral spat survival. *PLoS ONE* **2014**, *9*, e87545. [CrossRef] [PubMed]

38. Toh, T.C.; Ng, C.S.L.; Peh, J.W.K.; Toh, K.B.; Chou, L.M. Augmenting the post-transplantation growth and survivorship of juvenile scleractinian corals via nutritional enhancement. *PLoS ONE* **2014**, *9*, e98529. [CrossRef] [PubMed]

39. Horoszowski-Fridman, Y.B.; Brêthes, J.-C.; Rahmani, N.; Rinkevich, B. Marine silviculture: Incorporating ecosystem engineering properties into reef restoration acts. *Ecol. Eng.* **2015**, *82*, 201–213. [CrossRef]

40. Chamberland, V.F.; Petersen, D.; Guest, J.R.; Petersen, U.; Brittsan, M.; Vermeij, M.J. New seeding approach reduces costs and time to outplant sexually propagated corals for reef restoration. *Sci. Rep.* **2017**, *7*, 18076. [CrossRef] [PubMed]

41. Pollock, F.J.; Katz, S.M.; van de Water, J.A.; Davies, S.W.; Hein, M.; Torda, G.; Matz, M.V.; Beltran, V.H.; Buerger, P.; Puill-Stephan, E.; et al. Coral larvae for restoration and research: A large-scale method for rearing *Acropora millepora* larvae, inducing settlement, and establishing symbiosis. *Peer J.* **2017**, *5*, e3732. [CrossRef] [PubMed]

42. Tagliafico, A.; Rangel, S.; Christidis, L.; Kelaher, B.P. A potential method for improving coral self-attachment. *Restor. Ecol.* **2018**, *26*, 1082–1090. [CrossRef]

43. Tagliafico, A.; Rangel, S.; Kelaher, B.; Scheffers, S.; Christidis, L. A new technique to increase polyp production in stony coral aquaculture using waste fragments without polyps. *Aquaculture* **2018**, *484*, 303–308. [CrossRef]

44. Zayasu, Y.; Suzuki, G. Comparisons of population density and genetic diversity in artificial and wild populations of an arborescent coral, *Acropora yongei*: Implications for the efficacy of "artificial spawning hotspots". *Restor. Ecol.* **2019**, *27*, 440–446. [CrossRef]

45. Levi, G.; Shaish, L.; Haim, A.; Rinkevich, B. Mid-water rope nursery—Testing design and performance of a novel reef restoration instrument. *Ecol. Eng.* **2010**, *36*, 560–569. [CrossRef]

46. Horoszowski-Fridman, Y.B.; Izhaki, I.; Rinkevich, B. Engineering of coral reef larval supply through transplantation of nursery-farmed gravid colonies. *J. Exp. Mar. Biol. Ecol.* **2011**, *399*, 162–166. [CrossRef]

47. Horoszowski-Fridman, Y.B.; Rinkevich, B. Restoring the animal forests: Harnessing silviculture biodiversity concepts for coral transplantation. In *Marine Animal Forests: The Ecology of Benthic Biodiversity Hotspots*; Rossi, S., Bramanti, L., Gori, A., Orejas, C., Eds.; Springer International Publishing: Cham, Switzerland, 2017; pp. 1313–1335.

48. Sabater, M.G.; Yap, H.T. Growth and survival of coral transplants with and without electrochemical deposition of CaCO$_3$. *J. Exp. Mar. Biol. Ecol.* **2002**, *272*, 131–146. [CrossRef]

49. Borell, E.M.; Romatzki, S.B.C.; Ferse, S.C.A. Differential physiological responses of two congeneric scleractinian corals to mineral accretion and an electric field. *Coral Reef.* **2010**, *29*, 191–200. [CrossRef]

50. Strömberg, S.M.; Lundälv, T.; Goreau, T.J. Suitability of mineral accretion as a rehabilitation method for cold-water coral reefs. *J. Exp. Mar. Biol. Ecol.* **2010**, *395*, 153–161. [CrossRef]

51. Ng, C.S.L.; Toh, T.C.; Toh, K.B.; Guest, J.; Chou, L.M. Dietary habits of grazers influence their suitability as biological controls of fouling macroalgae in ex situ mariculture. *Aquac. Res.* **2014**, *45*, 1852–1860.

52. Villanueva, R.D.; Baria, M.V.B.; dela Cruz, D.W. Effects of grazing by herbivorous gastropod (*Trochus niloticus*) on the survivorship of cultured coral *Spat. Zool. Stud.* **2013**, *52*, 44. [CrossRef]

53. Forsman, Z.H.; Page, C.A.; Toonen, R.J.; Vaughan, D. Growing coral larger and faster: Micro-colony-fusion as a strategy for accelerating coral cover. *Peer J.* **2015**, *3*, e1313. [CrossRef] [PubMed]

54. Frias-Torres, S.; Van de Geer, C. Testing animal-assisted cleaning prior to transplantation in coral reef restoration. *Peer J.* **2015**, *3*, e1287. [CrossRef] [PubMed]

55. Chou, L.M. Rehabilitation engineering of Singapore reefs to cope with urbanization and climate change impacts. *JCEA* **2016**, *10*, 932–936.

56. Montoya-Maya, P.H.; Smit, K.P.; Burt, A.J.; Frias-Torres, S. Large-scale coral reef restoration could assist natural recovery in Seychelles, Indian Ocean. *J. Nat. Conserv.* **2016**, *16*, 1–17. [CrossRef]

57. Anthony, K.; Bay, L.K.; Costanza, R.; Firn, J.; Gunn, J.; Harrison, P.; Heyward, A.; Lundgren, P.; Mead, D.; Moore, T.; et al. New interventions are needed to save coral reefs. *Nat. Ecol. Evol.* **2017**, *1*, 1420–1422. [CrossRef] [PubMed]

58. Linden, B.; Rinkevich, B. Elaborating of an eco-engineering approach for stock enhanced sexually derived coral colonies. *J. Exp. Mar. Biol. Ecol.* **2017**, *486*, 314–321. [CrossRef]

59. Page, C.A.; Muller, E.M.; Vaughan, D.E. Microfragmenting for the successful restoration of slow growing massive corals. *Ecol. Eng.* **2018**, *123*, 86–94. [CrossRef]

60. Rachmilovitz, E.N.; Rinkevich, B. Tiling the reef—Exploring the first step of an ecological engineering tool that may promote phase-shift reversals in coral reefs. *Ecol. Eng.* **2017**, *105*, 150–161. [CrossRef]

61. Knoester, E.G.; Murk, A.J.; Osinga, R. Benefits of herbivorous fish outweigh costs of corallivory in coral nurseries placed close to a Kenyan patch reef. *Mar. Ecol. Prog. Ser.* **2019**, *611*, 143–155. [CrossRef]

62. Hoegh-Guldberg, O.; Hughes, L.; McIntyre, S.; Lindenmayer, D.B.; Parmesan, C.; Possingham, H.P.; Thomas, C.D. Assisted colonization and rapid climate change. *Science* **2008**, *321*, 345–346. [CrossRef] [PubMed]

63. Coles, S.L.; Riegl, B.M. Thermal tolerances of reef corals in the Gulf: A review of the potential for increasing coral survival and adaptation to climate change through assisted translocation. *Mar. Pollut. Bull.* **2013**, *72*, 323–332. [CrossRef] [PubMed]

64. Riegl, B.M.; Purkis, S.J.; Al-Cibahy, A.S.; Abdel-Moati, M.A.; Hoegh-Guldberg, O. Present limits to heat-adaptability in corals and population-level responses to climate extremes. *PLoS ONE* **2011**, *6*, e24802. [CrossRef]

65. Baker, A.C. Ecosystems—Reef corals bleach to survive change. *Nature* **2001**, *411*, 765–766. [CrossRef]

66. Schopmeyer, S.A.; Lirman, D.; Bartels, E.; Byrne, J.; Gilliam, D.S.; Hunt, J.; Johnson, M.E.; Larson, E.A.; Maxwell, K.; Nedimyer, K.; et al. In situ coral nurseries serve as genetic repositories for coral reef restoration after an extreme cold-water event. *Restor. Ecol.* **2012**, *20*, 696–703. [CrossRef]

67. Van Oppen, M.J.; Oliver, J.K.; Putnam, H.M.; Gates, R.D. Building coral reef resilience through assisted evolution. *Proc. Natl. Acad. Sci. USA* **2015**, *112*, 2307–2313. [CrossRef] [PubMed]

68. Bay, R.A.; Palumbi, S.R. Multilocus adaptation associated with heat resistance in reef-building corals. *Curr. Biol.* **2014**, *24*, 2952–2956. [CrossRef] [PubMed]

69. Chakravarti, L.J.; van Oppen, M.J.H. Experimental evolution in coral photosymbionts as a tool to increase thermal tolerance. *Front. Mar. Sci.* **2018**, *5*, 227. [CrossRef]

70. Chan, W.Y.; Peplow, L.M.; Menéndez, P.; Hoffmann, A.A.; van Oppen, M.J. Interspecific hybridization may provide novel opportunities for coral reef restoration. *Front. Mar. Sci.* **2018**, *5*, 160. [CrossRef]

71. Quigley, K.M.; Bay, L.K.; Willis, B.L. Leveraging new knowledge of *Symbiodinium* community regulation in corals for conservation and reef restoration. *Mar. Ecol. Prog. Ser.* **2018**, *600*, 245–253. [CrossRef]

72. Reshef, L.; Koren, O.; Loya, Y.; Zilber-Rosenberg, I.; Rosenberg, E. The coral probiotic hypothesis. *Environ. Microbiol.* **2006**, *8*, 2068–2073. [CrossRef] [PubMed]

73. Peixoto, R.S.; Rosado, P.M.; Leite, D.C.; Rosado, A.S. Beneficial microorganisms for corals (BMC): Proposed mechanisms for coral health and resilience. *Front. Microbiol.* **2017**, *8*, 341. [CrossRef] [PubMed]

74. Epstein, H.E.; Smith, H.A.; Torda, G.; van Oppen, M.J. Microbiome engineering: Enhancing climate resilience in corals. *Front. Ecol. Environ.* **2019**, *17*, 100–108. [CrossRef]

75. Rosado, P.M.; Leite, D.C.; Duarte, G.A.; Chaloub, R.M.; Jospin, G.; da Rocha, U.N.; Saraiva, J.P.; Dini-Andreote, F.; Eisen, J.A.; Bourne, D.G.; et al. Marine probiotics: Increasing coral resistance to bleaching through microbiome manipulation. *ISME J.* **2019**, *13*, 921–936. [CrossRef]

76. Potts, D.C. Natural-selection in experimental populations of reef-building corals (*Scleractinia*). *Evolution* **1984**, *38*, 1059–1078. [CrossRef]

77. Putnam, H.M.; Gates, R.D. Preconditioning in the reef-building coral *Pocillopora damicornis* and the potential for trans-generational acclimatization in coral larvae under future climate change conditions. *J. Exp. Biol.* **2015**, *8*, 2365–2372. [CrossRef] [PubMed]

78. Putnam, H.M.; Davidson, J.M.; Gates, R.D. Ocean acidification influences host DNA methylation and phenotypic plasticity in environmentally susceptible corals. *Evol. Appl.* **2016**, *9*, 1165–1178. [CrossRef] [PubMed]

79. McClanahan, T.R. Changes in coral sensitivity to thermal anomalies. *Mar. Ecol. Prog. Ser.* **2017**, *570*, 71–85. [CrossRef]

80. Dixon, G.B.; Bay, L.K.; Matz, M.V. Bimodal signatures of germline methylation are linked with gene expression plasticity in the coral *Acropora millepora*. *BMC Genomics* **2014**, *15*, 1109.

81. Palumbi, S.R.; Barshis, D.J.; Traylor-Knowles, N.; Bay, R.A. Mechanisms of reef coral resistance to future climate change. *Science* **2014**, *344*, 895–898. [CrossRef] [PubMed]

82. Dimond, J.L.; Gamblewood, S.K.; Roberts, S.B. Genetic and epigenetic insight into morphospecies in a reef coral. *Mol. Ecol.* **2017**, *26*, 5031–5042. [CrossRef] [PubMed]

83. Liew, Y.J.; Zoccola, D.; Li, Y.; Tambutté, E.; Venn, A.A.; Michell, C.T.; Cui, G.; Deutekom, E.S.; Kaandorp, J.A.; Voolstra, C.R.; et al. Epigenome-associated phenotypic acclimatization to ocean acidification in a reef-building coral. *Sci. Adv.* **2018**, *4*, eaar8028. [CrossRef] [PubMed]

84. Raymundo, L.J.; Maypa, A.P. Getting bigger faster: Mediation of size-specific mortality via fusion in juvenile coral transplants. *Ecol. Appl.* **2004**, *14*, 281–295. [CrossRef]

85. Rinkevich, B. Coral chimerism as an evolutionary rescue mechanism to mitigate global climate change impacts. *Glob. Change Biol.* **2019**, *25*, 1198–1206. [CrossRef] [PubMed]

86. Amar, K.O.; Chadwick, N.E.; Rinkevich, B. Coral kin aggregations exhibit mixed allogeneic reactions and enhanced fitness during early ontogeny. *BMC Evol. Biol.* **2008**, *8*, 126. [CrossRef] [PubMed]

87. Puill-Stephan, E.; van Oppen, M.J.H.; Pichavant-Rafini, K.; Willis, B.L. High potential for formation and persistence of chimeras following aggregated larval settlement in the broadcast spawning coral, *Acropora millepora*. *Proc. R. Soc. Lond. Ser. B Biol. Sci.* **2012**, *279*, 699–708. [CrossRef] [PubMed]

88. Mizrahi, D.; Navarrete, S.A.; Flores, A.A.V. Groups travel further: Pelagic metamorphosis and polyp clustering allow higher dispersal potential in sun coral propagules. *Coral Reef.* **2014**, *33*, 443–448. [CrossRef]

89. Toh, T.C.; Chou, L.M. Aggregated settlement of *Pocillopora damicornis* planulae on injury sites may facilitate coral wound healing. *Bull. Mar. Sci.* **2013**, *89*, 583–584. [CrossRef]

90. Rinkevich, B.; Shaish, L.; Douek, J.; Ben-Shlomo, R. Venturing in coral larval chimerism: A compact functional domain with fostered genotypic diversity. *Sci. Rep.* **2016**, *6*, 19493. [CrossRef] [PubMed]

91. Jiang, L.; Lei, X.M.; Liu, S.; Huang, H. Fused embryos and pre-metamorphic conjoined larvae in a broadcast spawning reef coral. *F1000 Res.* **2015**, *4*, 44. [CrossRef]

92. Mitsch, W. When will ecologists learn engineering and engineers learn ecology? *Ecol. Eng.* **2014**, *65*, 9–14. [CrossRef]

93. Wild, C.; Hoegh-Guldberg, O.; Naumann, M.S.; Colombo-Pallotta, M.F.; Ateweberhan, M.; Fitt, W.K.; Iglesias-Prieto, R.; Palmer, C.; Bythell, J.C.; Ortiz, J.C.; et al. Climate change impedes scleractinian corals as primary reef ecosystem engineers. *Mar. Freshw. Res.* **2011**, *62*, 205–215. [CrossRef]

94. Aronson, J.; Clewell, A.; Moreno-Mateos, D. Ecological restoration and ecological engineering: Complementary or indivisible? *Ecol. Eng.* **2016**, *91*, 392–395. [CrossRef]

95. Mitsch, W.J.; Jørgensen, S.E. *Ecological Engineering and Ecosystem Restoration*; John Wiley & Sons: Hoboken, NJ, USA, 2004.

96. Poore, A.G.; Graham, S.E.; Byrne, M.; Dworjanyn, S.A. Effects of ocean warming and lowered pH on algal growth and palatability to a grazing gastropod. *Mar. Biol.* **2016**, *163*, 99. [CrossRef]

97. Amar, K.O.; Rinkevich, B. A floating mid-water coral nursery as larval dispersion hub: Testing an idea. *Mar. Biol.* **2007**, *151*, 713–718. [CrossRef]

98. Shaish, L.; Levy, G.; Katzir, G.; Rinkevich, B. Coral reef restoration (Bolinao, Philippines) in the face of frequent natural catastrophes. *Restor. Ecol.* **2010**, *18*, 285–299. [CrossRef]

99. Precht, W.F.; Aronson, R.B. Climate flickers and range shifts of reef corals. *Front. Ecol. Environ.* **2004**, *2*, 307–314. [CrossRef]

100. Greenstein, B.J.; Pandolfi, J.M. Escaping the heat: Range shifts of reef coral taxa in coastal Western Australia. *Glob. Change Biol.* **2008**, *14*, 513–528. [CrossRef]

101. Yamano, H.; Sugihara, K.; Nomura, K. Rapid poleward range expansion of tropical reef corals in response to rising sea surface temperatures. *Geophys. Res. Lett.* **2011**, *38*, L04601. [CrossRef]

102. Denis, V.; Mezaki, T.; Tanaka, K.; Kuo, C.Y.; De Palmas, S.; Keshavmurthy, S.; Chen, C.A. Coverage, diversity, and functionality of a high-latitude coral community (Tatsukushi, Shikoku Island, Japan). *PLoS ONE* **2013**, *8*, e54330. [CrossRef] [PubMed]

103. Baird, A.H.; Sommer, B.; Madin, J.S. Pole-ward range expansion of *Acropora* spp. along the east coast of Australia. *Coral Reef.* **2012**, *31*, 1063. [CrossRef]

104. Tuckett, C.A.; de Bettignies, T.; Fromont, J.; Wernberg, T. Expansion of corals on temperate reefs: Direct and indirect effects of marine heatwaves. *Coral Reef.* **2017**, *36*, 947–956. [CrossRef]

105. Muir, P.R.; Wallace, C.C.; Done, T.; Aguirre, J.D. Limited scope for latitudinal extension of reef corals. *Science* **2015**, *348*, 1135–1138. [CrossRef] [PubMed]

106. Collyer, M.L.; Heilveil, J.S.; Stockwell, C.A. Contemporary evolutionary divergence for a protected species following assisted colonization. *PLoS ONE* **2011**, *6*, e22310. [CrossRef]

107. Chauvenet, A.L.M.; Ewen, J.G.; Armstrong, D.P.; Blackburn, T.M.; Pettorelli, N. Maximizing the success of assisted colonizations. *Anim. Conserv.* **2013**, *16*, 161–169. [CrossRef]

108. Wadgymar, S.M.; Cumming, M.N.; Weis, A.E. The success of assisted colonization and assisted gene flow depends on phenology. *Glob. Change Biol.* **2015**, *21*, 3786–3799. [CrossRef] [PubMed]

109. Kreyling, J.; Bittner, T.; Jaeschke, A.; Jentsch, A.; Jonas Steinbauer, M.; Thiel, D.; Beierkuhnlein, C. Assisted colonization: A question of focal units and recipient localities. *Restor. Ecol.* **2011**, *19*, 433–440. [CrossRef]

110. Jokiel, P.L. Long distance dispersal of reef corals by rafting. *Coral Reef.* **1984**, *3*, 113–116. [CrossRef]

111. Van Oppen, M.J.H.; Puill-Stephan, E.; Lundgren, P.; De'ath, G.; Bay, L.K. First-generation fitness consequences of interpopulational hybridisation in a Great Barrier Reef coral and its implications for assisted migration management. *Coral Reef.* **2014**, *33*, 607–611. [CrossRef]

112. Filbee-Dexter, K.; Smajdor, A. Ethics of assisted evolution in marine conservation. *Front. Mar. Sci.* **2019**, *6*, 20. [CrossRef]

113. Baums, I.B. A restoration genetics guide for coral reef conservation. *Mol. Ecol.* **2008**, *17*, 2796–2811. [CrossRef]

114. Raimundo, R.L.; Guimaraes, P.R., Jr.; Evans, D.M. Adaptive networks for restoration ecology. *Trends Ecol. Evol.* **2018**, *33*, 664–675. [CrossRef]

115. Levin, R.A.; Voolstra, C.R.; Agrawal, S.; Steinberg, P.D.; Suggett, D.J.; van Oppen, M.J.H. Engineering strategies to decode and enhance the genomes of coral symbionts. *Front. Microbiol.* **2017**, *8*, 1220. [CrossRef]

116. National Academies of Sciences, Engineering, and Medicine. *A Research Review of Interventions to Increase the Persistence and Resilience of Coral Reefs*; The National Academies Press: Washington, DC, USA, 2019.

117. Webster, N.S.; Reusch, T.B. Microbial contributions to the persistence of coral reefs. *ISME J.* **2017**, *11*, 2167–2174. [CrossRef]

118. Bošković, A.; Rando, O.J. Transgenerational epigenetic inheritance. *Annu. Rev. Genet.* **2018**, *52*, 21–41. [CrossRef]

119. Cropley, J.E.; Suter, C.M.; Beckman, K.B.; Martin, D.I. Germ-line epigenetic modification of the murine Avy allele by nutritional supplementation. *Proc. Natl. Acad. Sci. USA* **2006**, *103*, 17308–17312. [CrossRef] [PubMed]

120. Cooney, C.A.; Dave, A.A.; Wolff, G.L. Maternal methyl supplements in mice affect epigenetic variation and DNA methylation of offspring. *J. Nutr.* **2002**, *132*, S2393–S2400. [CrossRef] [PubMed]

121. Champagne, F.A.; Weaver, I.C.; Diorio, J.; Dymov, S.; Szyf, M.; Meaney, M.J. Maternal care associated with methylation of the estrogen receptor-α1b promoter and estrogen receptor-α expression in the medial preoptic area of female offspring. *Endocrinology* **2006**, *147*, 2909–2915. [CrossRef] [PubMed]

122. Bongiorni, L.; Giovanelli, D.; Rinkevich, B.; Pusceddu, A.; Chou, L.M.; Danovaro, R. First step in the restoration of a highly degraded coral reef (Singapore) by in situ coral intensive farming. *Aquaculture* **2011**, *322*, 191–200. [CrossRef]

123. Rinkevich, B. Will two walk together, except they have agreed? *J. Evol. Biol.* **2004**, *17*, 1178–1179. [CrossRef]

124. Rinkevich, B. Quo vadis chimerism? *Chimerism* **2011**, *2*, 1–5. [CrossRef] [PubMed]

125. Frank, U.; Oren, U.; Loya, Y.; Rinkevich, B. Alloimmune maturation in the coral *Stylophora pistillata* is achieved through three distinctive stages, four months post metamorphosis. *Proc. R. Soc. Lond. Ser. B Biol. Sci.* **1997**, *264*, 99–104. [CrossRef]

126. Barki, Y.; Gateño, D.; Graur, D.; Rinkevich, B. Soft-coral natural chimerism: A window in ontogeny allows the creation of entities comprised of incongruous parts. *Mar. Ecol. Prog. Ser.* **2002**, *23*, 91–99. [CrossRef]

127. SER (Society for Ecological). *The SER International Primer on Ecological Restoration*; Society for Ecological Restoration International: Tucson, AZ, USA, 2006.

128. Hobbs, R.J. Looking for the silver lining: Making the most of failure. *Restor. Ecol.* **2009**, *17*, 1–3.

129. Prober, S.M.; Doerr, V.A.; Broadhurst, L.M.; Williams, K.J.; Dickson, F. Shifting the conservation paradigm: A synthesis of options for renovating nature under climate change. *Ecol. Monog.* **2019**, *89*, e01333. [CrossRef]

130. Higgs, E.; Falk, D.A.; Guerrini, A.; Hall, M.; Harris, J.; Hobbs, R.J.; Jackson, S.T.; Rhemtulla, J.M.; Throop, W. The changing role of history in restoration ecology. *Front. Ecol. Environ.* **2014**, *12*, 499–506. [CrossRef]

131. O'Dea, R.E.; Noble, D.W.; Johnson, S.L.; Hesselson, D.; Nakagawa, S. The role of non-genetic inheritance in evolutionary rescue: Epigenetic buffering, heritable bet hedging and epigenetic traps. *Environ. Epigenet.* **2016**, *2*, dvv014. [CrossRef] [PubMed]

132. Williams, S.L.; Sur, C.; Janetski, N.; Hollarsmith, J.A.; Rapi, S.; Barron, L.; Heatwole, S.J.; Yusuf, A.M.; Yusuf, S.; Jompa, J.; et al. Large-scale coral reef rehabilitation after blast fishing in Indonesia. *Rest. Ecol.* **2019**, *27*, 447–456. [CrossRef]

133. Foo, S.A.; Asner, G.P. Scaling up coral reef restoration using remote sensing technology. *Front. Mar. Sci.* **2019**, *6*, 79. [CrossRef]

Journal of
*Marine Science
and Engineering*

Review

Thermal Stress and Resilience of Corals in a Climate-Changing World

Rodrigo Carballo-Bolaños [1,2,3], **Derek Soto** [1,2,3] **and Chaolun Allen Chen** [1,2,3,4,5,*]

1   Biodiversity Program, Taiwan International Graduate Program, Academia Sinica and National Taiwan
    Normal University, Taipei 11529, Taiwan; rodragoncar@gmail.com (R.C.-B.); dereksoto@gmail.com (D.S.)
2   Biodiversity Research Center, Academia Sinica, Taipei 11529, Taiwan
3   Department of Life Science, National Taiwan Normal University, Taipei 10610, Taiwan
4   Department of Life Science, Tunghai University, Taichung 40704, Taiwan
5   Institute of Oceanography, National Taiwan University, Taipei 10617, Taiwan
*   Correspondence: cac@gate.sinica.edu.tw

Received: 5 December 2019; Accepted: 20 December 2019; Published: 24 December 2019

Abstract: Coral reef ecosystems are under the direct threat of increasing atmospheric greenhouse gases, which increase seawater temperatures in the oceans and lead to bleaching events. Global bleaching events are becoming more frequent and stronger, and understanding how corals can tolerate and survive high-temperature stress should be accorded paramount priority. Here, we review evidence of the different mechanisms that corals employ to mitigate thermal stress, which include association with thermally tolerant endosymbionts, acclimatisation, and adaptation processes. These differences highlight the physiological diversity and complexity of symbiotic organisms, such as scleractinian corals, where each species (coral host and microbial endosymbionts) responds differently to thermal stress. We conclude by offering some insights into the future of coral reefs and examining the strategies scientists are leveraging to ensure the survival of this valuable ecosystem. Without a reduction in greenhouse gas emissions and a divergence from our societal dependence on fossil fuels, natural mechanisms possessed by corals might be insufficient towards ensuring the ecological functioning of coral reef ecosystems.

Keywords: thermal stress; coral resilience; bleaching events; thermally-tolerant symbionts; acclimatisation; adaptation; heterotrophy; climate change

---

1. Introduction

Since the last century, scleractinian coral reef ecosystems have undergone a decrease in biodiversity and ecological functioning [1–5], formerly attributed to the direct and indirect effects of overfishing [6,7], pollution from agriculture, sewage runoff, and land development [8–10]. Currently, along with the exponential increase of the human population [11] and our societal dependence on carbon fossil fuels, these local threats have been compounded by the impacts of global climate change in the oceans [12–14]. The impact of increasing greenhouse gases in the atmosphere is leading to a global increase in seawater temperatures that has caused mass bleaching events [12,14–17]. These global bleaching events are becoming more frequent (1998, 2010 and 2014–17) and severe [14,16,18–22], leaving coral reefs vulnerable and unable to recover. The 2014–2017 mass bleaching event, which lasted 36 months and spanned four calendar years, was the longest-lasting, most widespread, and probably most damaging event on record [21–29], and stands out as unique by spanning all phases of the El Niño-Southern Oscillation cycle of 2017, being the warmest non-El Niño year ever recorded [21,30].

Coral bleaching is defined as the loss of colour, due to the partial or total loss of Symbiodiniaceae dinoflagellates and/or the reduction of their photosynthetic pigments, that exposes the white calcium carbonate of the coral skeleton (Figure 1A) [31,32]. Bleaching is a generalized stress response to environmental perturbations such as aerial exposure, sedimentation, eutrophication, exposure to heavy metals, high UV radiation, and extreme changes in salinity and temperature [17,31,33,34], however, at large scales is triggered by high seawater temperatures (exceeding normal summer maxima) in combination with high solar radiation [12,15,17,31–33,35]. Scleractinian corals possess molecular protective mechanisms, such as heat shock proteins and antioxidant enzymes to resist thermal stress [17,33,36], or mycosporine amino acids (MAA) and fluorescent pigments to resist light stress (Figure 1B) [17,33,37,38]. The cellular mechanism of bleaching starts with the photoinhibition process within the photosynthetic apparatus of the endosymbionts, which results in the build-up of free electrons that react to form reactive oxygen species (ROS) [39,40]. The proliferation of harmful ROS leads to the degradation, exocytosis, or apoptosis of symbiont cells by the coral host [39], in order to avoid cellular damage [36]. If the duration of the thermal stress extends beyond their physiological ability to recover, corals cannot survive without their main symbiotic partners [15,31,35]. Even though the molecular process of bleaching is similar across coral species, variations in the mechanism to resist and survive thermal stress exist (Figure 1B,C) [17,32,35].

Resilience is the capacity of a coral colony or an entire coral reef ecosystem to absorb, resist, and recover from perturbations [41–43]. The resilience of corals to thermal stress is contingent on the mean long-term annual maximum temperature of the region they live in [17]. Much research has been done in the past decades to understand if the resilience of corals to thermal stress might be an adaptation and/or acclimatisation process (reviewed in [17,44,45]). Here, we review current research that focuses on the capabilities of coral species to adapt and/or acclimatise to thermal events, in order to understand what the future of this irreplaceable ecosystem will be. We have included only scientific studies which have clearly identified the different general strategies to survive thermal stress, as presented in this review.

2. Mechanisms of Resilience to Thermal Stress

2.1. Thermally Tolerant Endosymbionts

By associating with stress-resistant symbionts, some coral species are able to acquire increased thermal tolerance. Within the Symbiodiniaceae, species like *Durusdinium* spp. (previously clade D) [46–48], *Cladocopium* C15 [49], and *C. thermophilum* C3 [50,51] are resistant to thermal stress. Dinoflagellates in the genus *Durusdinium* are extremophiles inhabiting environments of thermal stress, high temperature fluctuations, sedimentation and high-latitudinal marginal reefs [52–62]. In recent decades, *Durusdinium* spp. have generated interest because they proliferate in bleached corals [53,60,63–65], protecting against thermal stress by providing 1–1.5 °C of thermal tolerance [46]. *Durusdinium* spp. maintain high photochemical efficiency when exposed to high temperatures compared to symbionts from other genera (*Breviolum* or *Cladocopium*) [48,66,67] and are able to fix more carbon and assimilate more nitrogen [68]. Furthermore, *D. trenchii* has been found to provide tolerance to cold stress too [69,70].

When exposed to thermal stress, some species of coral are capable of shifting the relative abundance of their dominant symbionts. Background symbionts, which can represent <10% of the overall Symbiodiniaceae community [60,71], become dominant, conferring thermal tolerance to the holobiont. Even though many coral species are able to associate with a heterogeneous community of Symbiodiniaceae [72,73], others do not change their dominant symbiont even when bleaching [74], showing a long-term symbiotic adaptation between coral host and dominant symbiont [75–78].

Figure 1. (**A**) Onset of a bleaching process in a colony of *Acropora* spp., in Kenting, Taiwan 2015 (Photo: J. Wei); (**B**) Bleaching event showing colonies with fluorescent pigments as a protective mechanism (**a**) and already bleached colonies (**b**), in Okinawa, Japan 2016 (Photo: S.-Y. Yang); (**C**) Intra-specific: between *Montipora* spp. colonies (**a**) and between *Isopora palifera* colonies (**b**), inter-specific: between *Montipora* spp. and *I. palifera* colonies (**c**) and intra-colony: within *Leptoria phrygia* colony (d) responses to thermal stress in Kenting, Taiwan 2016 (Photo: R. Carballo-Bolaños).

2.2. Acclimatisation (Phenotypic Plasticity)

Phenotypic plasticity refers to dissimilar phenotypes that can be generated from a single genotype in response to different environmental conditions [79]. These phenotypic changes are reversible and dependent on the boundaries of each organism's genotype [17]. In this context, acclimatisation refers to the phenotypic changes of corals in their natural environment, while acclimation denotes short-term phenotypic changes under manipulative experimental conditions in the laboratory. Reciprocal transplantation experiments (RTE) are a well-known method to quantify acclimatisation mechanisms by measuring differences in physiological parameters in specimens transplanted across environmentally distinct sites, locations or regions. For example, a RTE of *Porites lobata* between a fore reef (impacted by high wave action, oceanic swells and storms) and back reef (sheltered) in American Samoa, showed phenotypic plasticity in mean annual skeletal extension rates, bulk densities, and calcification rates after only six months, with all variables in transplanted corals approximating values of corals originally from the site [80]. In another study, Sawall et al. [81] found optimal calcification rates at 28–29 °C throughout all populations of *Pocillopora verrucosa* with evident differences in temperature fluctuations between the northern (21–27 °C) and southern (28–33 °C) parts of the Red Sea, supporting high phenotypic plasticity due to low genetic divergence between north and south coral host populations.

2.3. Thermal Stress Acclimatisation

Multiple studies have identified a direct link between thermal preconditioning and bleaching susceptibility (Table 1) [82–91]. After exposing corals to short-term thermal preconditioning experiments, only preconditioned corals did not bleach during a heat-stress experiment (Table 1) [82,83], despite maintaining their Symbiodiniaceae and the bacterial community [82]. Moreover, other studies have compared coral responses of the first major mass bleaching event in 1998 with subsequent stronger bleaching events [89,91]. Maynard et al. [91] surveyed the same sites in 1998 and after a more severe bleaching event in 2002, which featured exposure to twice as many degree heating weeks (DHW) and 15% higher solar irradiance, corals acclimatised, and exhibited less bleaching than in 1998. In a similar study, Guest et al. [89] demonstrated how coral bleaching was less severe after the 2010 large-scale bleaching event in Southeast Asia in locations that previously showed high bleaching in 1998 (Singapore and Malaysia), and had greater historical temperature variability and lower rates of warming. Meanwhile, corals in Indonesia were unaffected by bleaching in 1998, but showed high mortality in 2010. Consequently, corals acclimatised to previous thermal stress events, but also those living in sites with highly variable temperatures presented higher tolerance [89].

Table 1. Studies performed to test acclimatisation to thermal stress and high temperature variability at different locations around the world.

Study	Temp/DHW	Duration	Location	Species	Main Results	Ref.
Precondition + HSE	28 °C (precond.)	10 d	GBR, Australia	*Acropora millepora*	No bleaching in pre-conditioned corals	[82]
	31 °C (HSE)	8 d				
Precondition + HSE	31 °C (precond.)	2 d	GBR, Australia	*Acropora aspera*	No bleaching in pre-conditioned corals	[83]
	34 °C (HSE)	6 d				
Comparison 1998/2002 BE	DHW = 2002 > 1998	Bleaching survey	GBR, Australia	*Acropora* spp., *Pocillopora* spp., *Porites* spp.	Less mortality in 2002	[91]
Comparison 1998/2010 BE	DHW = Malaysia + Singapore > Indonesia	Bleaching survey	Indonesia, Malaysia, Singapore	*Acropora* spp., *Pocillopora* spp.,	Low bleaching in Malaysia and Singapore	[89]
Survey 2010 BE	-	Bleaching survey	Thailand	*Coelastrea aspera*	Less bleaching in high irradiance colony sides (decadal environmental 'memory')	[90]

Table 1. *Cont.*

Study	Temp/DHW	Duration	Location	Species	Main Results	Ref.
Comparison 1970/2017 HSE	31.4 °C	31 d	Hawaii, USA	*Montipora capitata, Pocillopora damicornis, Lobactis scutaria*	Higher calcification, delayed bleaching and mortality in 2017	[87]
HSE (HV and MV)	31.5 °C	5 d	American Samoa	*Acropora hyacinthus*	Mortality and photochemical efficiency decline: HV+*Durusdinium* < MV+*Durusdinium* < MV+*Cladocopium*	[67]
HSE from RT (HV and MV)	34 °C	3 h	American Samoa	*Acropora hyacinthus*	Acclimatised: MV to HV increased heat resistance; HV to MV reduced chl *a* retention; Different expression of 74 genes	[92]
HSE (HV and LV)	30 °C	270 d	Taiwan	*Pocillopora damicornis,*	Acclimatised: HV = control in all parameters	[93]
Comparison 1998/2005-06 BE (HV - LV)	-	Bleaching survey	Egypt, Madagascar, Seychelles, Australia, Guam, Kiribati, Cook Islands	Multiple species	Less bleaching in HV sites	[94]
Comparison multiple BEs (HV - LV)	-	Bleaching survey	Western Indian Ocean, Pacific Ocean, Caribbean Sea, GBR, Red Sea	Multiple species	Less bleaching in HV sites	[95]

HSE = Heat Stress Experiment, HV = Highly Variable, MV = Moderately Variable, LV = Low Variable, RT = Reciprocal Transplantation, BE = Bleaching Event, DHW = Degree Heating Weeks, d = days, h = hours, GBR = Great Barrier Reef.

Brown et al. [90] demonstrated 'long-term environmental memory' during the bleaching event in 2010. In 2000, coral colonies were rotated 180° in a manipulative experiment [96]. During the bleaching event of 2010, the sides of colonies exposed to high solar radiation before rotation in the 2000 experiment, retained four times as many symbionts than the sides exposed to low solar radiation, despite experiencing higher radiation for 10 years [90]. These experiments provide evidence that long-term acclimatisation to local conditions enhances thermal tolerance during bleaching events (Table 1). Coles et al. [87] showed evidence of acclimatisation to increasing seawater temperatures by replicating a bleaching experiment from 1970 at the same location in 2010. Because sea-surface temperature (SST) has steadily increased 1.13 °C over the last four decades, the authors experimentally increased 2.2 °C of ambient temperatures. Corals in 2017 showed higher calcification rates, delayed bleaching, and mortality compared to corals in 1970 (Table 1) [87]. Unfortunately, despite increased temperature tolerance in local corals, Hawaii suffered high coral mortality (34%) during the 2014–2017 global bleaching event, showing that high-temperature acclimatisation processes may not be occurring quickly enough to mitigate the projected length and intensity of future bleaching events [87].

2.4. Acclimatisation to High Temperature Variability

A series of backreef pools exhibiting tidal temperature variability on the island of Ofu, American Samoa, present a unique environment to study physiological differences between conspecific corals at small-spatial scales [97]. Using genetically identical coral fragments in a heat-stress experiment from both pools, Oliver and Palumbi [67] provided evidence of increased thermal tolerance when corals have acclimatised to high temperature variability (Table 1). Corals from the highly variable (HV) pool showed lower mortality and higher photochemical efficiency, while those from the moderately variable (MV) pool suffered increased mortality and lower photochemical efficiency related to symbiont species. Corals associated with *Durusdinium* spp. exhibited an intermediate decline in photochemical efficiency, while those associated with *Cladocopium* spp. showed the highest decline [67]. Palumbi et al. [92] performed reciprocal transplantations of corals between HV and MV pools and subjected those corals to a heat stress experiment to test for acclimatisation responses to thermal stress (Table 1). Corals acquired heat sensitivity based on the pool they were transplanted to: MV pool corals acquired heat resistance when moved to HV pool, but not to the same extent of HV conspecifics, while HV to MV transplantees experienced reduced chlorophyll *a* retention, similar to the levels of native corals [92]. Mayfield et al. [93] performed a thermal stress experiment with corals from a site in Taiwan exhibiting high daily temperature fluctuations and found that, under HV conditions, physiological parameters behaved similarly to those in control corals, suggesting that individuals living under HV temperatures can acclimate to high temperatures that would cause bleaching and mortality in unacclimated corals from other regions (Table 1) [93].

Some studies have compiled data of past bleaching events, in an effort to link patterns of bleaching susceptibility within sites under high temperature variability, in a worldwide context [94,95]. Sites characterized by a high-frequency pattern of temperature variability experienced higher thermal stress during both bleaching events, with extensive bleaching reported during 1998. However, in 2005–2006, these sites experienced reduced bleaching compared to sites under low frequency patterns, due to the acclimatisation of corals to thermal stress after the 1998 bleaching event and selective adaptation of resilient corals that survived the bleaching event [94]. Safaie et al. [95] explored this concept further by collecting in situ data with remotely sensed datasets from different reef locations around the globe, along with spatiotemporally coincident quantitative coral bleaching observations. Corals regularly exposed to temperature fluctuations on daily or tidal timescales became acclimatised to thermal stress and resistant to bleaching events. More importantly, these patterns of high-frequency temperature variability to bleaching occur in many reefs worldwide [95].

2.5. Molecular Mechanisms for Acclimatisation

Most studies involving transcriptomic analyses and thermal stress have shown differential gene expression under high temperature stress compared to controls (Table 2) [98–103]. Corals exposed to experimental thermal stress presented an upregulation of genes involved in oxidative stress responses [98,99] and carbon metabolism [98]. A comparison of differences in gene expression in corals preconditioned to thermal stress showed seventy differentially expressed genes between non-preconditioned corals and controls, 42 between preconditioned corals and controls, and nine genes between non-preconditioned and preconditioned corals (Table 2) [102].

Table 2. Thermal stress studies looking at differences in gene expression.

Study	Method	Genes/Molecules	Response	Location	Species	Main Results	Ref.
HSE – gene expression	qPCR	HSP70, MnSOD, ferritin, Zn^{2+}-metalloprotease	Oxidative stress response	GBR, Australia	*Acropora millepora*	Up-regulation in heat-stressed	[99]
HSE – gene expression	qPCR	HSP90, HSP70 Glyceraldehyde-3-phosphate dehydrogenase, α-ketoglutarate dehydrogenase, glycogen synthase, glycogen phosphorylase	Oxidative stress response Carbon metabolism	GBR, Australia	*Acropora aspera*	10.5-fold up-regulation in heat-stressed coral host genes	[98]
Short-term precondition—gene expression	cDNA micro-array	Lectins, heme-binding proteins, transcription factor AP-1, NF-kB inhibitor, thymosin, phosphate carrier protein, ferritin	Oxidative stress response	GBR, Australia	*Acropora aspera*	Different expression: 70 genes (non-preconditioned/control), 42 genes (preconditioned/control), 9 genes (preconditioned/non-preconditioned)	[102]
HSE—transcriptome wide gene expression (Thermally resilient vs. sensitive)	RNA-Seq	HSP70, HSP23/HSPB1, HSP16.2, CSMD1, Cu-Zn SOD TNFRs, TRAFs, NF-κB/Nfkb1, JNK/MAPK8	Oxidative stress response Apoptosis/immune response	American Samoa	*Acropora hyacinthus*	Up-regulation in thermal sensitive: 60 genes = "Frontloaded" in thermal resistant	[103]
HSE—transcriptome wide gene expression (RT HV – MV)	RNA-Seq	HSPs, Chaperonin proteins, CYPs TNFRs, TRAFs,	Oxidative stress response Apoptosis/immune response	American Samoa	*Acropora hyacinthus*	Different expression: 71 contigs based on pool of origin (HV/MV), 74 contigs based on pool transplanted	[93]

HSE = Heat Stress Experiment, HV = Highly Variable, MV = Moderately Variable, RT = Reciprocal Transplantation, qPCR = quantitative Polymerase Chain Reaction, HSP = Heat Stress Protein, SOD = Superoxide Dismutase, TNFR = Tumor Necrosis Factor Receptor, TRAF = Tumor Receptor Associated Family, GBR = Great Barrier Reef.

To understand the genomic basis of thermal resilience in corals, Barshis et al. [103] compared transcriptome-wide gene expression among thermally resilient and thermally sensitive conspecifics (Table 2). Sixty genes were up-regulated in thermally sensitive corals, while resilient corals already presented up-regulated genes under ambient conditions. These "frontloaded" genes facilitate a faster reaction to thermal stress at the protein level [103]. In a similar study, using reciprocally transplanted corals from HV and MV pools, transcriptome-wide gene expression analyses showed differential expression in 74 genes related to heat acclimation between genetically identical corals from both pools (Table 2) [93]. In a related study performed at the same sites, Ruiz-Jones and Palumbi [104] monitored the transcriptomic response of corals in the HV pool with a strong tidal cycle (high temperatures over 17 days). Their results bolstered the conclusions of Barshis et al. [103], showing that genes up-regulated during the hottest days, were enriched for "unfolded protein response", an ancient eukaryotic cellular response to endoplasmic reticulum stress, which corals use as the first line of defence against thermal stress [104].

2.6. Adaptation

Adaptation, strictly defined, refers to changes in the genetic composition of a population that are passed onto the next generation through natural selection [17,44,105]. The major concern regarding global climate change is that the current rate of environmental changes will outpace the evolutionary capabilities of corals to adapt [12,14,16,19,106]. Recent evidence has shown that, in addition to phenotypic plasticity and acclimatisation, other adaptive responses in corals, such as trans-generational plasticity [107], epigenetics [108,109], and somatic mutations [110] might contribute to resilience under thermal stress. Moreover, the fast rate of asexual reproduction within the Symbiodiniaceae (days to weeks *in hospite*) [111] in combination with large population sizes within corals ($\sim$1–5 $\times$ 10^6 cells cm^{-2}) [112] provide the potential for mutations to develop that might enable corals to resist thermal stress [110].

Few studies have examined adaptation to local thermal history in Symbiodiniaceae dinoflagellates [113,114]. Howells et al. [113] demonstrated adaptive capacity in the symbiont *C. goreaui* (formerly type C1) in corals from two sites in the GBR with dissimilar thermal histories. Corals hosting *C. goreaui* from the cooler site presented photodamage and bleaching, while those from the hotter site exhibited no signs of stress and greater growth [113]. Chakravarti et al. [114] tested adaptation to thermal tolerance of *C. goreaui* through experimental evolution. Dinoflagellates were cultured in vitro at elevated temperature of 31 °C for $\sim$80 generations (2.5 y), while wild-types were reared at 27 °C ambient temperature, then both cultures were tested at both temperatures. To measure physiological responses *in hospite*, both types (thermally selected and wild types) were inoculated into aposymbiotic recruits of three coral species and were exposed to both temperatures similar to in vitro experiments [114]. Symbionts reared in vitro performed better in photophysiology and growth at both temperatures, and showed lower levels of extracellular ROS. In contrast, wild-type symbionts were unable to photosynthesise or grow at high temperatures, and produced 17 times more extracellular ROS [114]. The differences were less obvious *in* hospite than in vitro. Cultures of corals inoculated with the thermally tolerant symbionts showed no difference in growth between 27 and 31 °C, while those inoculated with wild-types showed a negative growth trend at 31 °C, confirming an adaptation to thermal stress in *C. goreaui* after many generations living under high temperature [114].

Dixon et al. [115] revealed genetic data from the coral host that forms the heritable basis of temperature tolerance by performing a cross-fertilization experiment with coral colonies from two thermally divergent locations in GBR. The authors measured heat tolerance using the survivorship rate of larvae exposed to high temperatures and found that parents from the warmer location conferred significantly higher thermo-tolerance to their offspring, up to 10 fold increase in odds of survival, in comparison to parents from the cooler location. Dixon et al. [115] also identified "tolerance-associated genes" (TAGs), whose expression before stress predicted high survivorship rates in larvae under thermal stress, dissimilar from frontloaded genes [103]. When TAG expression was compared with parental colonies after three days of heat stress, they were negatively correlated with long-term heat stress response similar to the larval response, indicating that the larval heat tolerance results from the absence of pre-existing stress and not from prior up-regulation of heat stress genes through frontloading [115].

Krueger et al. [116] presented evidence that *Stylophora pistillata* underwent selection for heat tolerance in the Red Sea, after spending 47 days at 1–2 °C above their long-term summer maximum and showed an increase in primary productivity. Fine et al. [117] demonstrated how different corals species showed no signs of stress after exposure to 33 °C for four weeks and proposed that corals that colonised the Gulf of Aqaba after the last ice age had to cross exceedingly warm waters (>32 °C in the summer) at the entrance of the Red Sea, maintaining this adaptation to heat tolerance until the present day.

2.7. Heterotrophy

Heterotrophic carbon can become a significant energy source for some coral species when phototrophic carbon is unavailable, such as during a bleaching event (Table 1) [118]. Some studies have shown how heterotrophy replenished energy reserves in corals exposed to high temperatures [119] and during the recovery phase [120]. Similarly, Borell and Bischof [121] showed higher photochemical efficiency in fed corals compared to unfed corals after a mild thermal stress experiment. Also, Borell et al. [122] demonstrated how heterotrophy sustained photosynthetic activity and energy reserves in thermally stressed corals.

In a study which developed an energy-budget model linking coral bleaching and mortality risk, authors concluded that the time between the start of severe bleaching and the beginning of mortality is influenced by the amount of lipid stores corals have before the bleaching event and their capacity to acquire energy through heterotrophy [123]. With a stable isotope ^{13}C pulse-chase labelling experiment, Hughes et al. [124] demonstrated that, after exposure to high temperatures, coral hosts incorporated heterotrophic labelled carbon for storage and to stimulate endosymbiont recovery. Even after recovery from bleaching, 75% of carbon in newly acquired lipids was sourced heterotrophically [125], and corals continued assimilating heterotrophic carbon for up to 11 months after the bleaching experiment [126].

Nonetheless, the capacity for heterotrophic plasticity is compromised after two consecutive bleaching events [127]. Researchers experimentally bleached corals for 2.5 weeks, transferred corals to the field for recovery, and then repeated the bleaching experiment after one year. After the first thermal stress experiment, zooplankton and dissolved organic carbon (DOC) allowed the metabolic demand of bleached corals to be met; however, neither form of heterotrophic carbon was able to contribute to the energy budget of both species after the second bleaching experiment, suggesting that the capacity for heterotrophic plasticity is compromised under annual bleaching events [127] and corals need to depend on their energy reserves and/or symbiont association to survive repeated bleaching [128].

3. Perspectives for the Future

Because the loss of corals around the world would be a devastating consequence of human influence on earth, strategies to mitigate the damage and improve coral's thermal tolerance are currently being taken into consideration. For example, assisted colonization, migration and/or gene flow contemplate the movement of colonies or larvae of the same species living at different latitudes. 'Warm-adapted' corals can be transplanted to high latitude areas, where conspecifics living in colder environments, are vulnerable to thermal stress [129–131]. Assisted evolution has the potential to increase thermal stress tolerance in corals through various approaches: preconditioning acclimatisation (see Section: 'Thermal stress acclimatisation') and trans-generational acclimatisation, changes in microbial communities [132], selective breeding [133], mutagenesis [134], and the use of "CRISPR/Cas9" genome editing technology [135]. The use of 'strong corals' naturally adapted to high temperature extremes, such as corals originating from the Persian Gulf or Red Sea, as possible seedlings to repopulate areas where corals have disappeared [117,136–138] is also being considered. Unfortunately, none of these measures seem to be able to keep pace with the current rate of climate change, with the time between recurrent bleaching events becoming increasingly too short to allow complete recovery of coral reef ecosystems [139]. Despite recent advances in research methods and technology, such as transcriptomics [140], financial and logistical limitations to implement these actions remain [141], especially at large scales [130], and it takes many years to safely deploy new technology after social and political scrutiny [142].

Other conservation measures under consideration include designing better marine protected areas (MPAs) [143] or networks of MPAs [144–146], taking into consideration larval dispersal, connectivity and distribution patterns in areas with thermally tolerant corals [147] and including 'refugia' in areas where coral reefs have proven to be resilient to climate change [21,43,148,149]. This might help avoid the "protection paradox" in MPAs, in which vulnerable species are protected from local pressures, like fishing; yet while these species recover, they might be more sensitive to global pressures, like bleaching events [144]. Nevertheless, well-protected reefs within MPAs are not shielded from thermal stress [150,151]. After the last bleaching event, this was confirmed for MPAs [152], and for remote and isolated reefs with almost no direct human pressures [23,24,27,153–155].

The integration of assisted evolution [131,134] into coral reef restoration programs [156,157] to increase the resilience of already degraded ecosystems [41] is one strategy that has proven to be successful. Morikawa and Palumbi [158] used naturally thermal-tolerant corals from American Samoa to show that resilient corals can survive multiple bleaching events, providing the first proof that ecosystem engineering for conservation might be a resilience restoration tool of great importance in our climate changed future [158].

Evidence from reciprocally transplanted coral clones between sites with different thermal histories shows how individual coral colonies can shift their thermal threshold and thermal tolerance [93,159,160]. It is clear that many coral species are acclimatising and adapting to rapid changes in climate and their mechanisms differ among species and localities [67,82,83,89,90,113,115,161]. However, under current greenhouse gas emission projections, coral reefs worldwide are likely to change into new configurations with new assemblages of species [19,149,162–165]. These changes are happening fast, the GBR being the best example. After the 2014–2017 mass bleaching event, even the most 'pristine' areas in the northern GBR saw high mortality regardless of reefs' individual management status, proving that current management toolsets are insufficient to protect coral reef ecosystems from climate change [20,152]. The Paris Agreement was a first step to tackle the climate crisis, but no major industrialized country is meeting its pledges to control and reduce their greenhouse gas emissions [166]. It is imperative that societies completely change our dependence on fossil fuels, therefore addressing the root causes of climate change.

Author Contributions: Conceptualization, R.C.-B. and C.A.C.; methodology, R.C.-B.; software, R.C.-B.; validation, R.C.-B. and D.S.; formal analysis, R.C.-B.; investigation, R.C.-B.; resources, C.A.C.; data curation, R.C.-B.; writing—original draft preparation, R.C.-B. and D.S.; writing—review and editing, R.C.-B., D.S. and C.A.C.; visualization, R.C.-B.; supervision, C.A.C.; project administration, C.A.C.; funding acquisition, C.A.C. All authors have read and agreed to the published version of the manuscript.

Funding: This research was funded by Academia Sinica for Life Science Research (no. 4010) and Ministry of Science and Technology, Taiwan (MOST 101-2621-B-001-005-MY3) to C.A.C.

Acknowledgments: Many thanks to LK Chou and D Huang for their invitation to contribute to the special issue of "Coral Reef Resilience". We also thank three anonymous reviewers for helping improve this manuscript. R.C.-B. and D.S. are the receipts of PhD fellowship from Taiwan International Graduate Program (TIGP)-Biodiversity in the Academia Sinica. This review is supported by funds from Academia Sinica and Ministry of Science and Technology, Taiwan to C.A.C.

References

1. Wilkinson, C. *Status of Coral Reefs of the World: 2000*; Australian Institute of Marine Science: Townsville, Australia, 2000.
2. Gardner, T.A.; Côté, I.M.; Gill, J.A.; Grant, A.; Watkinson, A.R. Long-term region-wide declines in Caribbean corals. *Science* **2003**, *301*, 958–960. [CrossRef] [PubMed]
3. Pandolfi, J.M.; Bradbury, R.H.; Sala, E.; Hughes, T.P.; Bjorndal, K.A.; Cooke, R.G.; McArdle, D.; McClenachan, L.; Newman, M.J.H.; Paredes, G.; et al. Global trajectories of the long-term decline of coral reef ecosystems. *Science* **2003**, *301*, 955–958. [CrossRef] [PubMed]
4. De'ath, G.; Fabricius, K.E.; Sweatman, H.; Puotinen, M. The 27–year decline of coral cover on the Great Barrier Reef and its causes. *Proc. Natl. Acad. Sci. USA* **2012**, *109*, 17995–17999. [CrossRef] [PubMed]
5. Bruno, J.F.; Selig, E.R. Regional decline of coral cover in the Indo-Pacific: Timing, extent, and subregional comparisons. *PLoS ONE* **2007**, *2*, e711. [CrossRef]
6. Jackson, J.B.; Kirby, M.X.; Berger, W.H.; Bjorndal, K.A.; Botsford, L.W.; Bourque, B.J.; Bradbury, R.H.; Cooke, R.; Erlandson, J.; Estes, J.A.; et al. Historical overfishing and the recent collapse of coastal ecosystems. *Science* **2001**, *293*, 629–637. [CrossRef]
7. Valentine, J.F.; Heck, K.L. Perspective review of the impacts of overfishing on coral reef food web linkages. *Coral Reefs* **2005**, *24*, 209–213. [CrossRef]
8. Dubinsky, Z.; Stambler, N. Marine pollution and coral reefs. *Glob. Chang. Biol.* **1996**, *2*, 511–526. [CrossRef]
9. McCulloch, M.; Fallon, S.; Wyndham, T.; Hendy, E.; Lough, J.; Barnes, D. Coral record of increased sediment flux to the inner Great Barrier Reef since European settlement. *Nature* **2003**, *421*, 727. [CrossRef]
10. Fabricius, K.E. Effects of terrestrial runoff on the ecology of corals and coral reefs: Review and synthesis. *Mar. Pollut. Bull.* **2005**, *50*, 125–146. [CrossRef]
11. Cohen, J.E. Population growth and earth's human carrying capacity. *Science* **1995**, *269*, 341–346. [CrossRef]
12. Hughes, T.P.; Baird, A.H.; Bellwood, D.R.; Card, M.; Connolly, S.R.; Folke, C.; Grosberg, R.; Hoegh-Guldberg, O.; Jackson, J.B.C.; Kleypas, J.; et al. Climate change, human impacts, and the resilience of coral reefs. *Science* **2003**, *301*, 929–933. [CrossRef] [PubMed]
13. D'Angelo, C.; Wiedenmann, J. Impacts of nutrient enrichment on coral reefs: New perspectives and implications for coastal management and reef survival. *Curr. Opin. Environ. Sustain.* **2014**, *7*, 82–93. [CrossRef]
14. Hoegh-Guldberg, O.; Mumby, P.J.; Hooten, A.J.; Steneck, R.S.; Greenfield, P.; Gomez, E.; Harvell, C.D.; Sale, P.F.; Edwards, A.J.; Caldeira, K.; et al. Coral reefs under rapid climate change and ocean acidification. *Science* **2007**, *318*, 1737–1742. [CrossRef] [PubMed]
15. Fitt, W.K.; Brown, B.E.; Warner, M.E.; Dunne, R.P. Coral bleaching: Interpretation of thermal tolerance limits and thermal thresholds in tropical corals. *Coral Reefs* **2001**, *20*, 51–65. [CrossRef]
16. Hoegh-Guldberg, O. Climate change, coral bleaching and the future of the world's coral reefs. *Mar. Freshw. Res.* **1999**, *50*, 839–866. [CrossRef]
17. Coles, S.L.; Brown, B.E. Coral bleaching—Capacity for acclimatization and adaptation. *Adv. Mar. Biol.* **2003**, *46*, 183–223. [CrossRef]
18. Heron, S.F.; Maynard, J.A.; Van Hooidonk, R.; Eakin, C.M. Warming trends and bleaching stress of the world's coral reefs 1985–2012. *Sci. Rep.* **2016**, *6*, 38402. [CrossRef]
19. Hughes, T.P.; Barnes, M.L.; Bellwood, D.R.; Cinner, J.E.; Cumming, G.S.; Jackson, J.B.; Kleypas, J.; van der Leemput, I.A.; Lough, J.M.; Morrison, T.H.; et al. Coral reefs in the Anthropocene. *Nature* **2017**, *546*, 82. [CrossRef]
20. Hughes, T.P.; Kerry, J.T.; Álvarez-Noriega, M.; Álvarez-Romero, J.G.; Anderson, K.D.; Baird, A.H.; Babcock, R.C.; Beger, M.; Bellwood, D.R.; Berkelmans, R.; et al. Global warming and recurrent mass bleaching of corals. *Nature* **2017**, *543*, 373–377. [CrossRef]
21. Eakin, C.M.; Sweatman, H.P.A.; Brainard, R.E. The 2014–2017 global-scale coral bleaching event: Insights and impacts. *Coral Reefs* **2019**, *38*, 539–545. [CrossRef]
22. Skirving, W.J.; Heron, S.F.; Marsh, B.L.; Liu, G.; De La Cour, J.L.; Geiger, E.F.; Eakin, C.M. The relentless march of mass coral bleaching: A global perspective of changing heat stress. *Coral Reefs* **2019**, *38*, 547–557. [CrossRef]

23. Harrison, H.B.; Álvarez-Noriega, M.; Baird, A.H.; Heron, S.F.; MacDonald, C.; Hughes, T.P. Back-to-back coral bleaching events on isolated atolls in the Coral Sea. *Coral Reefs* **2019**, *38*, 713–719. [CrossRef]

24. Head, C.E.I.; Bayley, D.T.I.; Rowlands, G.; Roche, R.C.; Tickler, D.M.; Rogers, A.D.; Koldewey, H.; Turner, J.R.; Andradi-Brown, D.A. Coral bleaching impacts from back-to-back 2015–2016 thermal anomalies in the remote central Indian Ocean. *Coral Reefs* **2019**, *38*, 605–618. [CrossRef]

25. Raymundo, L.J.; Burdick, D.; Hoot, W.C.; Miller, R.M.; Brown, V.; Reynolds, T.; Gault, J.; Idechong, J.; Fifer, J.; Williams, A. Successive bleaching events cause mass coral mortality in Guam, Micronesia. *Coral Reefs* **2019**, *38*, 677–700. [CrossRef]

26. Burt, J.A.; Paparella, F.; Al-Mansoori, N.; Al-Mansoori, A.; Al-Jailani, H. Causes and consequences of the 2017 coral bleaching event in the southern Persian/Arabian Gulf. *Coral Reefs* **2019**, *38*, 567–589. [CrossRef]

27. Vargas-Ángel, B.; Huntington, B.; Brainard, R.E.; Venegas, R.; Oliver, T.; Barkley, H.; Cohen, A. El Niño-associated catastrophic coral mortality at Jarvis Island, central Equatorial Pacific. *Coral Reefs* **2019**, *38*, 731–741. [CrossRef]

28. Frade, P.R.; Bongaerts, P.; Englebert, N.; Rogers, A.; Gonzalez-Rivero, M.; Hoegh-Guldberg, O. Deep reefs of the Great Barrier Reef offer limited thermal refuge during mass coral bleaching. *Nat. Commun.* **2018**, *9*, 3447. [CrossRef]

29. Ku'ulei, S.R.; Bahr, K.D.; Jokiel, P.L.; Donà, A.R. Patterns of bleaching and mortality following widespread warming events in 2014 and 2015 at the Hanauma Bay Nature Preserve, Hawaii. *Peer J.* **2017**, *5*, e3355.

30. Hartfield, G.; Blunden, J.; Arndt, D.S. State of the climate in 2017. *Bull. Am. Meteorol. Soc.* **2018**, *99*, S1–S310. [CrossRef]

31. Glynn, P. Coral reef bleaching: Ecological perspectives. *Coral Reefs* **1993**, *12*, 1–17. [CrossRef]

32. Douglas, A.E. Coral bleaching—how and why? *Mar. Pollut. Bull.* **2003**, *46*, 385–392. [CrossRef]

33. Brown, B.E. Coral bleaching: Causes and consequences. *Coral Reefs* **1997**, *16*, S129–S138. [CrossRef]

34. Obura, D.O. Reef corals bleach to resist stress. *Mar. Pollut. Bull.* **2009**, *58*, 206–212. [CrossRef] [PubMed]

35. Jokiel, P.L. Temperature stress and coral bleaching. In *Coral Health and Disease*; Springer: Berlin, Germany, 2004; pp. 401–425.

36. Lesser, M.P. Oxidative stress in marine environments: Biochemistry and physiological ecology. *Annu. Rev. Physiol.* **2006**, *68*, 253–278. [CrossRef]

37. Shick, J.M.; Dunlap, W.C. Mycosporine-like amino acids and related gadusols: Biosynthesis, accumulation, and UV-protective functions in aquatic organisms. *Annu. Rev. Physiol.* **2002**, *64*, 223–262. [CrossRef]

38. Salih, A.; Larkum, A.; Cox, G.; Kühl, M.; Hoegh-Guldberg, O. Fluorescent pigments in corals are photoprotective. *Nature* **2000**, *408*, 850–853. [CrossRef]

39. Weis, V.M. Cellular mechanisms of cnidarian bleaching: Stress causes the collapse of symbiosis. *J. Exp. Biol.* **2008**, *211*, 3059–3066. [CrossRef]

40. Tchernov, D.; Gorbunov, M.Y.; de Vargas, C.; Yadav, S.N.; Milligan, A.J.; Häggblom, M.; Falkowski, P.G. Membrane lipids of symbiotic algae are diagnostic of sensitivity to thermal bleaching in corals. *Proc. Natl. Acad. Sci. USA* **2004**, *101*, 13521–13535. [CrossRef]

41. Darling, E.S.; Côté, I.M. Seeking resilience in marine ecosystems. *Science* **2018**, *359*, 986–987. [CrossRef]

42. Bellwood, D.R.; Hughes, T.P.; Folke, C.; Nyström, M. Confronting the coral reef crisis. *Nature* **2004**, *429*, 827–833. [CrossRef]

43. Roche, R.C.; Williams, G.J.; Turner, J.R. Towards developing a mechanistic understanding of coral reef resilience to thermal stress across multiple scales. *Curr. Clim. Chang. Rep.* **2018**, *4*, 51–64. [CrossRef]

44. Edmunds, P.J.; Gates, R.D. Acclimatization in tropical reef corals. *Mar. Ecol. Prog. Ser.* **2008**, *361*, 307–310. [CrossRef]

45. Brown, B.E.; Cossins, A.R. The potential for temperature acclimatisation of reef corals in the face of climate change. In *Coral Reefs: An Ecosystem in Transition*; Springer: Berlin, Germany, 2011; pp. 421–433.

46. Berkelmans, R.; van Oppen, M.J.H. The role of zooxanthellae in the thermal tolerance of corals: A 'nugget of hope' for coral reefs in an era of climate change. *Proc. R. Soc. B Biol. Sci.* **2006**, *273*, 2305–2312. [CrossRef] [PubMed]

47. Stat, M.; Gates, R.D. Clade D *Symbiodinium* in scleractinian corals: A "nugget" of hope, a selfish opportunist, an ominous sign, or all of the above? *J. Mar. Biol.* **2011**, *2011*. [CrossRef]

48. Rowan, R. Thermal adaptation in reef coral symbionts. *Nature* **2004**, *430*, 742. [CrossRef] [PubMed]

49. Levas, S.J.; Grottoli, A.G.; Hughes, A.; Osburn, C.L.; Matsui, Y. Physiological and biogeochemical traits of bleaching and recovery in the mounding species of coral *Porites lobata*: Implications for resilience in mounding corals. *PLoS ONE* **2013**, *8*, e63267. [CrossRef]

50. Hume, B.C.C.; Voolstra, C.R.; Arif, C.; D'Angelo, C.; Burt, J.A.; Eyal, G.; Loya, J.; Wiedenmann, J. Ancestral genetic diversity associated with the rapid spread of stress-tolerant coral symbionts in response to Holocene climate change. *Proc. Natl. Acad. Sci. USA* **2016**, *113*, 4416–4421. [CrossRef]

51. Hume, B.C.; D'Angelo, C.; Smith, E.G.; Stevens, J.R.; Burt, J.; Wiedenmann, J. *Symbiodinium thermophilum* sp. nov., a thermotolerant symbiotic alga prevalent in corals of the world's hottest sea, the Persian/Arabian Gulf. *Sci. Rep.* **2015**, *5*, 8562. [CrossRef]

52. Chen, C.A.; Wang, J.-T.; Fang, L.-S.; Yang, Y.-W. Fluctuating algal symbiont communities in *Acropora palifera* (Scleractinia: Acroporidae) in Taiwan. *Mar. Ecol. Prog. Ser.* **2005**, *295*, 113–121. [CrossRef]

53. Hsu, C.-M.; Keshavmurthy, S.; Dennis, V.; Kuo, C.-Y.; Wang, J.-T.; Meng, P.-J.; Chen, C.A. Temporal and spatial variations of symbiont communities in catch bowl coral, *Isopora palifera* (Scleractinia; Acroporidae), at reefs in Kenting National Park, Taiwan. *Zool. Stud.* **2012**, *51*, 1343–1353.

54. LaJeunesse, T.C.; Pettay, D.T.; Sampayo, E.M.; Phongsuwan, N.; Brown, B.; Obura, D.O.; Hoegh-Guldberg, O.; Fitt, W.K. Long-standing environmental conditions, geographic isolation and host-symbiont specificity influence the relative ecological dominance and genetic diversification of coral endosymbionts in the genus *Symbiodinium*. *J. Biogeogr.* **2010**, *37*, 785–800. [CrossRef]

55. Lien, Y.-T.; Nakano, Y.; Plathong, S.; Fukami, H.; Wang, J.-T.; Chen, C.A. Occurrence of the putatively heat-tolerant *Symbiodinium* phylotype D in high-latitudinal outlying coral communities. *Coral Reefs* **2007**, *26*, 35–44. [CrossRef]

56. Wham, D.C.; Ning, G.; LaJeunesse, T.C. *Symbiodinium glynnii* sp. nov., a species of stress-tolerant symbiotic dinoflagellates from pocilloporid and montiporid corals in the Pacific Ocean. *Phycologia* **2017**, *56*, 396–409. [CrossRef]

57. LaJeunesse, T.C.; Parkinson, J.E.; Gabrielson, P.W.; Jeong, H.J.; Reimer, J.D.; Voolstra, C.R.; Santos, S.R. Systematic revision of Symbiodiniaceae highlights the antiquity and diversity of coral endosymbionts. *Curr. Biol.* **2018**, *28*, 2570–2580. [CrossRef] [PubMed]

58. Carballo-Bolaños, R.; Denis, V.; Huang, Y.-Y.; Keshavmurthy, S.; Chen, C.A. Temporal variation and photochemical efficiency of species in Symbiodiniaceae associated with coral *Leptoria phrygia* (Scleractinia; Merulinidae) exposed to contrasting temperature regimes. *PLoS ONE* **2019**, *14*, e0218801. [CrossRef] [PubMed]

59. Chen, C.A.; Lam, K.K.; Nakano, Y.; Tsai, W.-S. A stable association of the stress-tolerant zooxanthellae, *Symbiodinium* clade D, with the low-temperature-tolerant coral, *Oulastrea crispata* (Scleractinia: Faviidae) in subtropical non-reefal coral communities. *Zool. Stud.* **2003**, *42*, 540–550.

60. Kao, K.-W.; Keshavmurthy, S.; Tsao, C.-H.; Wang, J.-T.; Chen, C.A. Repeated and prolonged temperature anomalies negate Symbiodiniaceae genera shuffling in the coral *Platygyra verweyi* (Scleractinia; Merulinidae). *Zool. Stud.* **2018**, *57*, 1–14.

61. Keshavmurthy, S.; Hsu, C.-M.; Kuo, C.-Y.; Meng, P.-J.; Wang, J.-T.; Chen, C.A. Symbiont communities and host genetic structure of the brain coral *Platygyra verweyi*, at the outlet of a nuclear power plant and adjacent areas. *Mol. Ecol.* **2012**, *21*, 4393–4407. [CrossRef]

62. Keshavmurthy, S.; Meng, P.-J.; Wang, J.-T.; Kuo, C.-Y.; Yang, S.-Y.; Hsu, C.-M.; Gan, C.-H.; Dai, C.-F.; Chen, C.A. Can resistant coral-*Symbiodinium* associations enable coral communities to survive climate change? A study of a site exposed to long-term hot water input. *Peer J.* **2014**, *2*, e327. [CrossRef]

63. Lajeunesse, T.C.; Smith, R.T.; Finney, J.; Oxenford, H. Outbreak and persistence of opportunistic symbiotic dinoflagellates during the 2005 Caribbean mass coral 'bleaching' event. *Proc. R. Soc. B Biol. Sci.* **2009**, *276*, 4139–4148. [CrossRef]

64. Silverstein, R.N.; Cunning, R.; Baker, A.C. Change in algal symbiont communities after bleaching, not prior heat exposure, increases heat tolerance of reef corals. *Glob. Chang. Biol.* **2015**, *21*, 236–249. [CrossRef] [PubMed]

65. Jones, A.M.; Berkelmans, R.; van Oppen, M.J.H.; Mieog, J.C.; Sinclair, W. A community change in the algal endosymbionts of a scleractinian coral following a natural bleaching event: Field evidence of acclimatization. *Proc. R. Soc. B Biol. Sci.* **2008**, *275*, 1359–1365. [CrossRef] [PubMed]

66. Cunning, R.; Silverstein, R.N.; Baker, A.C. Symbiont shuffling linked to differential photochemical dynamics of *Symbiodinium* in three Caribbean reef corals. *Coral Reefs* **2017**, 1–8. [CrossRef]

67. Oliver, T.A.; Palumbi, S.R. Do fluctuating temperature environments elevate coral thermal tolerance? *Coral Reefs* **2011**, *30*, 429–440. [CrossRef]

68. Baker, D.M.; Andras, J.P.; Jordán-Garza, A.G.; Fogel, M.L. Nitrate competition in a coral symbiosis varies with temperature among Symbiodinium clades. *ISME J.* **2013**, *7*, 1248–1251. [CrossRef]

69. Silverstein, R.N.; Cunning, R.; Baker, A.C. Tenacious D: *Symbiodinium* in clade D remain in reef corals at both high and low temperature extremes despite impairment. *J. Exp. Biol.* **2017**, *220*, 1192–1196. [CrossRef]

70. LaJeunesse, T.C.; Smith, R.; Walther, M.; Pinzón, J.; Pettay, D.T.; McGinley, M.; Aschaffenburg, M.; Medina-Rosas, P.; Cupul-Magaña, A.L.; López Pérez, A.; et al. Host-symbiont recombination versus natural selection in the response of coral-dinoflagellate symbioses to environmental disturbance. *Proc. R. Soc. B Biol. Sci.* **2010**, *277*, 2925–2934. [CrossRef]

71. Mieog, J.C.; van Oppen, M.J.; Cantin, N.E.; Stam, W.T.; Olsen, J.L. Real-time PCR reveals a high incidence of *Symbiodinium* clade D at low levels in four scleractinian corals across the Great Barrier Reef: Implications for symbiont shuffling. *Coral Reefs* **2007**, *26*, 449–457. [CrossRef]

72. Silverstein, R.N.; Correa, A.M.S.; Baker, A.C. Specificity is rarely absolute in coral–algal symbiosis: Implications for coral response to climate change. *Proc. R. Soc. B Biol. Sci.* **2012**, *279*, 2609–2618. [CrossRef]

73. Baker, A.C. Flexibility and specificity in coral-agal symbiosis: Diversity, ecology, and biogeogrpahy of *Symbiodinium*. *Annu. Rev. Ecol. Evol. Syst.* **2003**, *34*, 661–689. [CrossRef]

74. Goulet, T.L. Most corals may not change their symbionts. *Mar. Ecol. Prog. Ser.* **2006**, *321*, 1–7. [CrossRef]

75. Thornhill, D.J.; Xiang, Y.; Fitt, W.K.; Santos, S.R. Reef endemism, host specificity and temporal stability in populations of symbiotic dinoflagellates from two ecologically dominant Caribbean corals. *PLoS ONE* **2009**, *4*, e6262. [CrossRef] [PubMed]

76. Stat, M.; Loh, W.K.W.; LaJeunesse, T.C.; Hoegh-Guldberg, O.; Carter, D.A. Stability of coral-endosymbiont associations during and after a thermal stress event in the southern Great Barrier Reef. *Coral Reefs* **2009**, *28*, 709–713. [CrossRef]

77. Thornhill, D.J.; Fitt, W.K.; Schmidt, G.W. Highly stable symbioses among western Atlantic brooding corals. *Coral Reefs* **2006**, *25*, 515–519. [CrossRef]

78. Thornhill, D.J.; Lajeunesse, T.C.; Kemp, D.W.; Fitt, W.K.; Schmidt, G.W. Multi-year, seasonal genotypic surveys of coral-algal symbioses reveal prevalent stability or post-bleaching reversion. *Mar. Biol.* **2006**, *148*, 711–722. [CrossRef]

79. Pigliucci, M.; Murren, C.J.; Schlichting, C.D. Phenotypic plasticity and evolution by genetic assimilation. *J. Exp. Biol.* **2006**, *209*, 2362–2367. [CrossRef]

80. Smith, L.W.; Barshis, D.; Birkeland, C. Phenotypic plasticity for skeletal growth, density, and calification of *Porites lobata* in response to habitat type. *Coral Reefs* **2007**, *26*, 559–567. [CrossRef]

81. Sawall, Y.; Al-Sofyani, A.; Hohn, S.; Banguera-Hinestroza, E.; Voolstra, C.R.; Wahl, M. Extensive phenotypic plasticity of a Red Sea coral over a strong latitudinal temperature gradient suggests limited acclimatization potential to warming. *Sci. Rep.* **2015**, *5*, 8940. [CrossRef]

82. Bellantuono, A.J.; Hoegh-Guldberg, O.; Rodriguez-Lanetty, M. Resistance to thermal stress in corals without changes in symbiont composition. *Proc. R. Soc. B Biol. Sci.* **2012**, *279*, 1100–1107. [CrossRef]

83. Middlebrook, R.; Hoegh-Guldberg, O.; Leggat, W. The effect of thermal history on the susceptibility of reef-building corals to thermal stress. *J. Exp. Biol.* **2008**, *211*, 1050–1056. [CrossRef]

84. Hawkins, T.D.; Warner, M.E. Warm preconditioning protects against acute heat-induced respiratory dysfunction and delays bleaching in a symbiotic sea anemone. *J. Exp. Biol.* **2017**, *220*, 969–983. [CrossRef] [PubMed]

85. Putnam, H.M.; Gates, R.D. Preconditioning in the reef-building coral *Pocillopora damicornis* and the potential for trans-generational acclimatization in coral larvae under future climate change conditions. *J. Exp. Biol.* **2015**, *218*, 2365–2372. [CrossRef] [PubMed]

86. Fisch, J.; Drury, C.; Towle, E.K.; Winter, R.N.; Miller, M.W. Physiological and reproductive repercussions of consecutive summer bleaching events of the threatened Caribbean coral *Orbicella faveolata*. *Coral Reefs* **2019**, *38*, 863–876. [CrossRef]

87. Coles, S.L.; Bahr, K.D.; Rodgers, K.u.S.; May, S.L.; McGowan, A.E.; Tsang, A.; Bumgarner, J.; Han, J.H. Evidence of acclimatization or adaptation in Hawaiian corals to higher ocean temperatures. *Peer J.* **2018**, *6*, e5347. [CrossRef] [PubMed]

88. Hughes, T.P.; Kerry, J.T.; Connolly, S.R.; Baird, A.H.; Eakin, C.M.; Heron, S.F.; Hoey, A.S.; Hoogenboom, M.O.; Jacobson, M.; Liu, G.; et al. Ecological memory modifies the cumulative impact of recurrent climate extremes. *Nat. Clim. Chang.* **2019**, *9*, 40–43. [CrossRef]

89. Guest, J.R.; Baird, A.H.; Maynard, J.A.; Muttaqin, E.; Edwards, A.J.; Campbell, S.J.; Yewdall, K.; Affendi, Y.A.; Chou, L.M. Contrasting patterns of coral bleaching susceptibility in 2010 suggest an adaptive response to thermal stress. *PLoS ONE* **2012**, *7*, e33353. [CrossRef]

90. Brown, B.E.; Dunne, R.P.; Edwards, A.J.; Sweet, M.J.; Phongsuwan, N. Decadal environmental 'memory' in a reef coral? *Mar. Biol.* **2015**, *162*, 479–483. [CrossRef]

91. Maynard, J.A.; Anthony, K.R.N.; Marshall, P.A.; Masiri, I. Major bleaching events can lead to increased thermal tolerance in corals. *Mar. Biol.* **2008**, *155*, 173–182. [CrossRef]

92. Palumbi, S.R.; Barshis, D.J.; Traylor-Knowles, N.; Bay, R.A. Mechanisms of reef coral resistance to future climate change. *Science* **2014**, *344*, 895–898. [CrossRef]

93. Mayfield, A.B.; Fan, T.-Y.; Chen, C.-S. Physiological acclimation to elevated temperature in a reef-building coral from an upwelling environment. *Coral Reefs* **2013**, *32*, 909–921. [CrossRef]

94. Thompson, D.M.; Van Woesik, R. Corals escape bleaching in regions that recently and historically experienced frequent thermal stress. *Proc. R. Soc. B Biol. Sci.* **2009**, *276*, 2893–2901. [CrossRef] [PubMed]

95. Safaie, A.; Silbiger, N.J.; McClanahan, T.R.; Pawlak, G.; Barshis, D.J.; Hench, J.L.; Rogers, J.S.; Williams, G.J.; Davis, K.A. High frequency temperature variability reduces the risk of coral bleaching. *Nat. Commun.* **2018**, *9*, 1671. [CrossRef] [PubMed]

96. Brown, B.; Dunne, R.; Warner, M.; Ambarsari, I.; Fitt, W.; Gibb, S.; Cummings, D.G. Damage and recovery of Photosystem II during a manipulative field experiment on solar bleaching in the coral *Goniastrea aspera*. *Mar. Ecol. Prog. Ser.* **2000**, *195*, 117–124. [CrossRef]

97. Thomas, L.; Rose, N.H.; Bay, R.A.; López, E.H.; Morikawa, M.K.; Ruiz-Jones, L.; Palumbi, S.R. Mechanisms of thermal tolerance in reef-building corals across a fine-grained environmental mosaic: Lessons from Ofu, American Samoa. *Front. Mar. Sci.* **2018**, *4*. [CrossRef]

98. Leggat, W.; Seneca, F.; Wasmund, K.; Ukani, L.; Yellowlees, D.; Ainsworth, T.D. Differential responses of the coral host and their algal symbiont to thermal stress. *PLoS ONE* **2011**, *6*, e26687. [CrossRef] [PubMed]

99. Császár, N.; Seneca, F.; Van Oppen, M. Variation in antioxidant gene expression in the scleractinian coral *Acropora millepora* under laboratory thermal stress. *Mar. Ecol. Prog. Ser.* **2009**, *392*, 93–102. [CrossRef]

100. DeSalvo, M.K.; Sunagawa, S.; Voolstra, C.R.; Medina, M. Transcriptomic responses to heat stress and bleaching in the elkhorn coral *Acropora palmata*. *Mar. Ecol. Prog. Ser.* **2010**, *402*, 97–113. [CrossRef]

101. Seneca, F.O.; Palumbi, S.R. The role of transcriptome resilience in resistance of corals to bleaching. *Mol. Ecol.* **2015**, *24*, 1467–1484. [CrossRef]

102. Bellantuono, A.J.; Granados-Cifuentes, C.; Miller, D.J.; Hoegh-Guldberg, O.; Rodriguez-Lanetty, M. Coral thermal tolerance: Tuning gene expression to resist thermal stress. *PLoS ONE* **2012**, *7*, e50685. [CrossRef]

103. Barshis, D.J.; Ladner, J.T.; Oliver, T.A.; Seneca, F.O.; Traylor-Knowles, N.; Palumbi, S.R. Genomic basis for coral resilience to climate change. *Proc. Natl. Acad. Sci. USA* **2013**, *110*, 1387–1392. [CrossRef]

104. Ruiz-Jones, L.J.; Palumbi, S.R. Tidal heat pulses on a reef trigger a fine-tuned transcriptional response in corals to maintain homeostasis. *Sci. Adv.* **2017**, *3*, e1601298. [CrossRef] [PubMed]

105. Brown, B.E. Adaptations of reef corals to physical environmental stress. *Adv. Mar. Biol.* **1997**, *31*, 222–299.

106. Hoegh-Guldberg, O. The adaptation of coral reefs to climate change: Is the Red Queen being outpaced? *Sci. Mar.* **2012**, *76*, 403–408. [CrossRef]

107. Torda, G.; Donelson, J.M.; Aranda, M.; Barshis, D.J.; Bay, L.; Berumen, M.L.; Bourne, D.G.; Cantin, N.; Foret, S.; Matz, M.; et al. Rapid adaptive responses to climate change in corals. *Nat. Clim. Chang.* **2017**, *7*, 627. [CrossRef]

108. Eirin-Lopez, J.M.; Putnam, H.M. Marine environmental epigenetics. *Annu. Rev. Mar. Sci.* **2019**, *11*, 335–368. [CrossRef] [PubMed]

109. Durante, M.K.; Baums, I.B.; Williams, D.E.; Vohsen, S.; Kemp, D.W. What drives phenotypic divergence among coral clonemates of *Acropora palmata*? *Mol. Ecol.* **2019**, *28*. [CrossRef]

110. Van Oppen, M.J.H.; Souter, P.; Howells, E.J.; Heyward, A.; Berkelmans, R. Novel genetic diversity through somatic mutations: Fuel for adaptation of reef corals? *Diversity* **2011**, *3*, 405–423. [CrossRef]

111. Wilkerson, F.; Kobayashi, D.; Muscatine, L. Mitotic index and size of symbiotic algae in Caribbean reef corals. *Coral Reefs* **1988**, *7*, 29–36. [CrossRef]

112. Fitt, W.K.; McFarland, F.K.; Warner, M.E.; Chilcoat, G.C. Seasonal patterns of tissue biomass and densities of symbiotic dinoflagellates in reef corals and relation to coral bleaching. *Limnol. Oceanogr.* **2000**, *45*, 677–685. [CrossRef]

113. Howells, E.J.; Beltrán, V.H.; Larsen, N.W.; Bay, L.K.; Willis, B.L.; van Oppen, M.J.H. Coral thermal tolerance shaped by local adaptation of photosymbionts. *Nat. Clim. Chang.* **2012**, *2*, 116–120. [CrossRef]

114. Chakravarti, L.J.; Beltran, V.H.; van Oppen, M.J. Rapid thermal adaptation in photosymbionts of reef-building corals. *Glob. Chang. Biol.* **2017**, *23*, 4675–4688. [CrossRef] [PubMed]

115. Dixon, G.B.; Davies, S.W.; Aglyamova, G.V.; Meyer, E.; Bay, L.K.; Matz, M.V. Genomic determinants of coral heat tolerance across latitudes. *Science* **2015**, *348*, 1460–1462. [CrossRef] [PubMed]

116. Krueger, T.; Horwitz, N.; Bodin, J.; Giovani, M.-E.; Escrig, S.; Meibom, A.; Fine, M. Common reef-building coral in the northern Red Sea resistant to elevated temperature and acidification. *R. Soc. Open Sci.* **2017**, *4*, 170038. [CrossRef] [PubMed]

117. Fine, M.; Gildor, H.; Genin, A. A coral reef refuge in the Red Sea. *Glob. Chang. Biol.* **2013**, *19*, 3640–3647. [CrossRef] [PubMed]

118. Houlbrèque, F.; Ferrierriers, C. Heterotrophy in tropical scleractinian corals. *Biol. Rev.* **2009**, *84*, 1–17. [CrossRef] [PubMed]

119. Grottoli, A.G.; Rodrigues, L.J.; Palardy, J.E. Heterotrophic plasticity and resilience in bleached corals. *Nature* **2006**, *440*, 1186–1189. [CrossRef]

120. Rodrigues, L.J.; Grottoli, A.G. Energy reserves and metabolism as indicators of coral recovery from bleaching. *Limnol. Oceanogr.* **2007**, *52*, 1874–1882. [CrossRef]

121. Borell, E.M.; Bischof, K. Feeding sustains photosythetic quantum yield of a scleractinian coral during thermal stress. *Oecologia* **2008**, *157*, 593–601. [CrossRef]

122. Borell, E.M.; Yuliantri, A.R.; Bischof, K.; Richter, C. The effect of heterotrophy on photosynthesis and tissue composition of two scleractinian corals under elevated temperature. *J. Exp. Mar. Biol. Ecol.* **2008**, *364*, 116–123. [CrossRef]

123. Anthony, K.; Hoogenboom, M.O.; Maynard, J.A.; Grottoli, A.G.; Middlebrook, R. Energetics approach to predicting mortality risk from environmental stress: A case study of coral bleaching. *Funct. Ecol.* **2009**, *23*, 539–550. [CrossRef]

124. Hughes, A.; Grottoli, A.; Pease, T.; Matsui, Y. Acquisition and assimilation of carbon in non-bleached and bleached corals. *Mar. Ecol. Prog. Ser.* **2010**, *420*, 91–101. [CrossRef]

125. Baumann, J.; Grottoli, A.G.; Hughes, A.D.; Matsui, Y. Photoautotrophic and heterotrophic carbon in bleached and non-bleached coral lipid acquisition and storage. *J. Exp. Mar. Biol. Ecol.* **2014**, *461*, 469–478. [CrossRef]

126. Hughes, A.D.; Grottoli, A.G. Heterotrophic compensation: A possible mechanism for resilience of coral reefs to global warming or a sign of prolonged stress? *PLoS ONE* **2013**, *8*, e81172. [CrossRef] [PubMed]

127. Levas, S.; Grottoli, A.G.; Schoepf, V.; Aschaffenburg, M.; Baumann, J.; Bauer, J.E.; Warner, M.E. Can heterotrophic uptake of dissolved organic carbon and zooplankton mitigate carbon budget deficits in annually bleached corals? *Coral Reefs* **2016**, *35*, 495–506. [CrossRef]

128. Grottoli, A.G.; Warner, M.E.; Levas, S.J.; Aschaffenburg, M.D.; Schoepf, V.; McGinley, M.; Baumann, J.; Matsui, Y. The cumulative impact of annual coral bleaching can turn some coral species winners into losers. *Glob. Chang. Biol.* **2014**, *20*, 3823–3833. [CrossRef]

129. Van Oppen, M.J.H.; Puill-Stephan, E.; Lundgren, P.; De'ath, G.; Bay, L.K. First-generation fitness consequences of interpopulational hybridisation in a Great Barrier Reef coral and its implications for assisted migration management. *Coral Reefs* **2014**, *33*, 607–611. [CrossRef]

130. Hoegh-Guldberg, O.; Hughes, L.; McIntyre, S.; Lindenmayer, D.B.; Parmesan, C.; Possingham, H.P.; Thomas, C.D. Assisted colonization and rapid climate change. *Science* **2008**, *321*, 345–346. [CrossRef]

131. Anthony, K.; Bay, L.K.; Costanza, R.; Firn, J.; Gunn, J.; Harrison, P.; Heyward, A.; Lundgren, P.; Mead, D.; Moore, T.; et al. New interventions are needed to save coral reefs. *Nat. Ecol. Evol.* **2017**, *1*, 1420. [CrossRef]

132. Webster, N.S.; Reusch, T.B.H. Microbial contributions to the persistence of coral reefs. *ISME J.* **2017**, *11*, 2167. [CrossRef]

133. Chan, W.Y.; Peplow, L.M.; Menéndez, P.; Hoffmann, A.A.; van Oppen, M.J.H. Interspecific hybridization may provide novel opportunities for coral reef restoration. *Front. Mar. Sci.* **2018**, *5*. [CrossRef]

134. Van Oppen, M.J.; Oliver, J.K.; Putnam, H.M.; Gates, R.D. Building coral reef resilience through assisted evolution. *Proc. Natl. Acad. Sci. USA* **2015**, *112*, 2307–2313. [CrossRef] [PubMed]

135. Levin, R.A.; Voolstra, C.R.; Agrawal, S.; Steinberg, P.D.; Suggett, D.J.; van Oppen, M.J.H. Engineering strategies to decode and enhance the genomes of coral symbionts. *Front. Microbiol.* **2017**, *8*. [CrossRef] [PubMed]

136. Spalding, M.D.; Brown, B.E. Warm-water coral reefs and climate change. *Science* **2015**, *350*, 769–771. [CrossRef] [PubMed]

137. Coles, S.L.; Riegl, B.M. Thermal tolerances of reef corals in the Gulf: A review of the potential for increasing coral survival and adaptation to climate change through assisted translocation. *Mar. Pollut. Bull.* **2013**, *72*, 323–332. [CrossRef]

138. Riegl, B.M.; Purkis, S.J.; Al-Cibahy, A.S.; Abdel-Moati, M.A.; Hoegh-Guldberg, O. Present limits to heat-adaptability in corals and population-level responses to climate extremes. *PLoS ONE* **2011**, *6*, e24802. [CrossRef]

139. Hughes, T.P.; Anderson, K.D.; Connolly, S.R.; Heron, S.F.; Kerry, J.T.; Lough, J.M.; Baird, A.H.; Baum, J.K.; Berumen, M.L.; Bridge, T.C.; et al. Spatial and temporal patterns of mass bleaching of corals in the Anthropocene. *Science* **2018**, *359*, 80–83. [CrossRef]

140. Cziesielski, M.J.; Schmidt-Roach, S.; Aranda, M. The past, present, and future of coral heat stress studies. *Ecol. Evol.* **2019**, *9*, 10055–10066. [CrossRef]

141. Bayraktarov, E.; Saunders, M.I.; Abdullah, S.; Mills, M.; Beher, J.; Possingham, H.P.; Mumby, P.J.; Lovelock, C.E. The cost and feasibility of marine coastal restoration. *Ecol. Appl.* **2016**, *26*, 1055–1074. [CrossRef]

142. Kaebnick, G.E.; Heitman, E.; Collins, J.P.; Delborne, J.A.; Landis, W.G.; Sawyer, K.; Taneyhill, L.A.; Winickoff, D.E. Precaution and governance of emerging technologies. *Science* **2016**, *354*, 710–711. [CrossRef]

143. Edgar, G.J.; Stuart-Smith, R.D.; Willis, T.J.; Kininmonth, S.; Baker, S.C.; Banks, S.; Barret, N.S.; Becerro, M.A.; Bernard, A.T.F.; Berkhout, J.; et al. Global conservation outcomes depend on marine protected areas with five key features. *Nature* **2014**, *506*, 216. [CrossRef]

144. Bates, A.E.; Cooke, R.S.C.; Duncan, M.I.; Edgar, G.J.; Bruno, J.F.; Benedetti-Cecchi, L.; Côté, I.M.; Lefcheck, J.S.; Costello, M.J.; Barrett, N.; et al. Climate resilience in marine protected areas and the 'Protection Paradox'. *Biol. Conserv.* **2019**, *236*, 305–314. [CrossRef]

145. Grorud-Colvert, K.; Claudet, J.; Tissot, B.N.; Caselle, J.E.; Carr, M.H.; Day, J.C.; Friedlander, A.M.; Lester, S.E.; De Loma, T.L.; Malone, D.; et al. Marine protected area networks: Assessing whether the whole is greater than the sum of its parts. *PLoS ONE* **2014**, *9*, e102298. [CrossRef] [PubMed]

146. Bellwood, D.R.; Pratchett, M.S.; Morrison, T.H.; Gurney, G.G.; Hughes, T.P.; Álvarez-Romero, J.G.; Day, J.C.; Grantham, R.; Grech, A.; Hoey, A.S.; et al. Coral reef conservation in the Anthropocene: Confronting spatial mismatches and prioritizing functions. *Biol. Conserv.* **2019**, *236*, 604–615. [CrossRef]

147. Mumby, P.J.; Elliott, I.A.; Eakin, C.M.; Skirving, W.; Paris, C.B.; Edwards, H.J.; Enríquez, S.; Iglesias-Prieto, R.; Cherubin, L.M.; Stevens, J.R. Reserve design for uncertain responses of coral reefs to climate change. *Ecol. Lett.* **2011**, *14*, 132–140. [CrossRef] [PubMed]

148. Beyer, H.L.; Kennedy, E.V.; Beger, M.; Chen, C.A.; Cinner, J.E.; Darling, E.S.; Eakin, C.M.; Gates, R.D.; Heron, S.F.; Knowlton, N.; et al. Risk-sensitive planning for conserving coral reefs under rapid climate change. *Conserv. Lett.* **2018**, *11*, e12587. [CrossRef]

149. Hoegh-Guldberg, O.; Kennedy, E.V.; Beyer, H.L.; McClennen, C.; Possingham, H.P. Securing a long-term future for coral reefs. *Trends Ecol. Evol.* **2018**, *33*, 936–944. [CrossRef]

150. Selig, E.R.; Casey, K.S.; Bruno, J.F. Temperature-driven coral decline: The role of marine protected areas. *Glob. Chang. Biol.* **2012**, *18*, 1561–1570. [CrossRef]

151. Bruno, J.F.; Côté, I.M.; Toth, L.T. Climate change, coral loss, and the curious case of the parrotfish paradigm: Why don't marine protected areas improve reef resilience? *Annu. Rev. Mar. Sci.* **2019**, *11*, 307–334. [CrossRef]

152. Hughes, T.P.; Kerry, J.T.; Baird, A.H.; Connolly, S.R.; Dietzel, A.; Eakin, C.M.; Heron, S.F.; Hoey, A.S.; Hoogenboom, M.O.; Liu, G.; et al. Global warming transforms coral reef assemblages. *Nature* **2018**, *556*, 492. [CrossRef]

153. Sheppard, C.; Sheppard, A.; Mogg, A.; Bayley, D.; Dempsey, A.C.; Roche, R.; Turner, J.; Purkis, S. Coral bleaching and mortality in the Chagos Archipelago. *Atoll Res. Bull.* **2017**, *613*, 1–26. [CrossRef]

154. Barkley, H.C.; Cohen, A.L.; Mollica, N.R.; Brainard, R.E.; Rivera, H.E.; DeCarlo, T.M.; Lohmann, G.P.; Drenkard, E.J.; Alpert, A.E.; Young, C.W.; et al. Repeat bleaching of a central Pacific coral reef over the past six decades (1960–2016). *Commun. Biol.* **2018**, *1*, 177. [CrossRef] [PubMed]

155. Fox, M.D.; Carter, A.L.; Edwards, C.B.; Takeshita, Y.; Johnson, M.D.; Petrovic, V.; Amir, C.G.; Sala, E.; Sandin, S.A.; Smith, J.E. Limited coral mortality following acute thermal stress and widespread bleaching on Palmyra Atoll, central Pacific. *Coral Reefs* **2019**, *38*, 701–712. [CrossRef]

156. Van Oppen, M.J.; Gates, R.D.; Blackall, L.L.; Cantin, N.; Chakravarti, L.J.; Chan, W.Y.; Cormick, C.; Crean, A.; Damjanovic, K.; Epstein, H.; et al. Shifting paradigms in restoration of the world's coral reefs. *Glob. Chang. Biol.* **2017**, *23*, 3437–3448. [CrossRef] [PubMed]

157. Rinkevich, B. The active reef restoration toolbox is a vehicle for coral resilience and adaptation in a changing world. *J. Mar. Sci. Eng.* **2019**, *7*, 201. [CrossRef]

158. Morikawa, M.K.; Palumbi, S.R. Using naturally occurring climate resilient corals to construct bleaching-resistant nurseries. *Proc. Natl. Acad. Sci. USA* **2019**, *116*, 10586–10591. [CrossRef] [PubMed]

159. Kenkel, C.D.; Matz, M.V. Gene expression plasticity as a mechanism of coral adaptation to a variable environment. *Nat. Ecol. Evol.* **2017**, *1*, 0014. [CrossRef] [PubMed]

160. Bay, R.A.; Palumbi, S.R. Transcriptome predictors of coral survival and growth in a highly variable environment. *Ecol. Evol.* **2017**, *7*, 4794–4803. [CrossRef]

161. Bay, R.A.; Rose, N.H.; Logan, C.A.; Palumbi, S.R. Genomic models predict successful coral adaptation if future ocean warming rates are reduced. *Sci. Adv.* **2017**, *3*, e1701413. [CrossRef]

162. Frieler, K.; Meinshausen, M.; Golly, A.; Mengel, M.; Lebek, K.; Donner, S.D.; Hoegh-Guldberg, O. Limiting global warming to 2 °C is unlikely to save most coral reefs. *Nat. Clim. Chang.* **2012**, *3*, 165. [CrossRef]

163. Van Hooidonk, R.; Maynard, J.; Tamelander, J.; Gove, J.; Ahmadia, G.; Raymundo, L.; Williams, G.; Heron, S.F.; Planes, S. Local-scale projections of coral reef futures and implications of the Paris Agreement. *Sci. Rep.* **2016**, *6*, 39666. [CrossRef]

164. Pachauri, R.K.; Allen, M.R.; Barros, V.R.; Broome, J.; Cramer, W.; Christ, R.; Church, J.A.; Clarke, L.; Dahe, Q.; Dasgupta, P.; et al. Climate change 2014: Synthesis report. In *Contribution of Working Groups I, II and III to the Fifth Assessment Report of the Intergovernmental Panel on Climate Change*; Ipcc: Geneva, Switzerland, 2014.

165. Graham, N.A.; Bellwood, D.R.; Cinner, J.E.; Hughes, T.P.; Norström, A.V.; Nyström, M. Managing resilience to reverse phase shifts in coral reefs. *Front. Ecol. Environ.* **2013**, *11*, 541–548. [CrossRef]

166. Spash, C.L. This changes nothing: The Paris Agreement to ignore reality. *Globalizations* **2016**, *13*, 928–933. [CrossRef]

Journal of
Marine Science and Engineering

Article

Coral Resilience at Malauka'a Fringing Reef, Kāne'ohe Bay, O'ahu after 18 years

Kelsey A. Barnhill [1,*] and Keisha D. Bahr [2]

[1] Faculty of Environmental Sciences and Natural Resources, Norwegian University of Life Sciences, P.O. Box 5003, 1432 Ås, Norway

[2] Department of Life Sciences, Texas A&M University-Corpus Christi, Corpus Christi, TX 78412, USA

* Correspondence: kelseybarnhill@gmail.com

Received: 18 July 2019; Accepted: 3 September 2019; Published: 6 September 2019

Abstract: Globally, coral reefs are under threat from climate change and increasingly frequent bleaching events. However, corals in Kāne'ohe Bay, Hawai'i have demonstrated the ability to acclimatize and resist increasing temperatures. Benthic cover (i.e., coral, algae, other) was compared over an 18 year period (2000 vs. 2018) to estimate species composition changes. Despite a climate change induced 0.96 °C temperature increase and two major bleaching events within the 18-year period, the fringing reef saw no significant change in total coral cover (%) or relative coral species composition in the two dominant reef-building corals, *Porites compressa* and *Montipora capitata*. However, the loss of two coral species (*Pocillopora meandrina* and *Porites lobata*) and the addition of one new coral species (*Leptastrea purpurea*) between surveys indicates that while the fringing reef remains intact, a shift in species composition has occurred. While total non-coral substrate cover (%) increased from 2000 to 2018, two species of algae (*Gracilaria salicornia* and *Kappaphycus alvarezii*) present in the original survey were absent in 2018. The previously dominant algae *Dictyosphaeria* spp. significantly decreased in percent cover between surveys. The survival of the studied fringing reef indicates resilience and suggests these Hawaiian corals are capable of acclimatization to climate change and bleaching events.

Keywords: coral reefs; macroalgae; resilience; species composition

---

1. Introduction

Warming sea surface temperatures caused by climate change threaten coral reefs globally [1]. Increased water temperatures cause coral bleaching (reviewed in Reference [2]) which can cause total or partial mortality for colonies if the corals are unable to recover (reviewed in Reference [3]). Coral mortality leads to reef degradation as the reef loses structural complexity and is overgrown by algae, often leading to an algae-dominated phase shift [4]. Reef degradation directly causes the loss of reef-related ecosystem services, such as seafood production, shoreline protection, habitat provision, materials for medicines, and nitrogen fixation, among others [5].

Significant ecological declines driven by anthropogenic stressors are occurring on coral reefs around the world [6]. In 2000, an estimated 11% of all coral reefs had already been lost with an additional 16% damaged beyond the point of being functional ecosystems [7]. From 1985–2012 the Great Barrier Reef experienced a 50.7% decrease in coral cover [6] and the coral cover in the entire Indo-Pacific is 20% less than historical levels from 100 years ago [8]. Hawaiian reefs, however, have one of the lowest threat ratings in the Pacific (less than 30% threatened) [9]. From 1999–2012 mean Hawaiian coral cover and diversity remained stable statewide, including within Kāne'ohe Bay [10]. Reefs within Kāne'ohe Bay have repeatedly shown resilience by recovering from natural and anthropogenic disturbances such as bleaching events [11]. Increasingly frequent bleaching events threaten the longevity of coral reef ecosystems [12] and whether or not corals can become adaptive or resistant to bleaching is contested in current literature [12]. However, corals in Kāne'ohe Bay have shown resilience through acclimatization to increased temperatures [13]. In this study resilience is defined as 'the ability of an ecosystem to recuperate its structure and functions after a perturbation' [14].

Kāne'ohe Bay, Hawai'i

Coral reefs in Kāne'ohe Bay, located on the northeast side of O'ahu, Hawai'i (21°4′ N and 157°8′ W), have some of the highest levels of coral cover (54–68% compared to statewide average of 24.1%) across the Hawaiian islands [10,11,15]. Due to the unique geographic properties of Kāne'ohe Bay, these reefs experience elevated summer water temperatures (1–2 °C), which offshore reefs will not be subjected to for several years [16].

Kāne'ohe Bay represents one of the few recorded examples of a phase shift reversal, in which the reefs were coral-dominant then algal-dominant and have returned to coral-dominated reefs all within a 40-year period [17]. From 1960–1970 the human population in Kāne'ohe doubled, leading to effluent municipal and military sewage to be discharged in the bay, causing eutrophication and a subsequent decline in coral cover and diversity [18]. Following the release of effluent sewage into the bay, the algae *Dictyosphaeria cavernosa*, stimulated by increased nutrient availability, spread widely, causing a phase shift from coral-dominated to algae-dominated [19,20]. Following the 1979 sewage diversion, coral cover in the bay more than doubled in just four years [21] as nutrient levels decreased [19].

The first documented coral bleaching event in Kāne'ohe occurred in 1996, in which the total coral mortality was < 1% [22]. A second, more severe bleaching event occurred in 2014 [16]. While nearly half of all corals in the southern region of the bay were pale or bleached immediately following a 2014 bleaching, there was only 1% total coral mortality three months later [23]. In 2015, another widespread bleaching event affected the Kāne'ohe Bay reefs, however a 15% decrease in bleaching compared to the 2014 event suggested some corals may be acclimatizing to increased temperatures, although higher levels of mortality were observed [11]. Kāne'ohe Bay has retained high coral cover despite Hawaiian offshore water temperatures increasing by 1.15 °C over the past 60 years [11]. Corals within the bay also show increased resistance to acidification and warming waters compared to other corals in O'ahu [24]. The historical resilience of corals in Kāne'ohe Bay and the consistently high coral cover while many reefs around the globe are in decline led to the following research question: How has coral cover and community composition changed in response to 18 years of warming temperatures and two major bleaching events in a well-studied coral reef ecosystem?

2. Materials and Methods

2.1. Study Site: Kāne'ohe Bay, Hawai'i

The study site was a 600-meter section of the Malauka'a fringing reef (21.44300899° N, 157.80636° W to 21.43853104° N, 157.806541° W) in the south-west of Kāne'ohe Bay, which was initially surveyed in 2000 [25]. Similar to other reefs in the bay, *Porites compressa* and *Montipora capitata* are the dominant reef-building corals. The northern section of the reef is approximately 125 meters offshore of Kealohi Point at He'eia State park. The southern 200 meters of the study site is adjacent to the Paepae o He'eia (traditional Hawaiian fishpond) where there is ongoing estuarine restoration focusing on sociocultural benefits [26]. The southern end of the reef is subjected to freshwater stream and pond output from He'eia stream and a triple mākāhā (sluice gate) within Paepae o He'eia [27]. The selected reef suffered bleaching and low mortality (<5%) during the 2014/2015 bleaching event [11].

2.2. Comparative Study Setup

2.2.1. Benthic Survey

Coral cover and benthic community composition were measured through a quali-quantitative comparison using a modified version of the point intercept transect (PIT) (as described by Reference [28]) in the initial survey (2000) and follow up survey (2018). The PIT method identifies benthic cover every 50 cm along a transect [29]. During the 2000 study [25], benthic cover was recorded every meter and thus repeated as such in the 2018 study. Coral species, algae species, crustose coralline algae, turf, sand, and rubble were recorded along each transect. Crustose coralline algae and turf were pooled together into 'non-coral substrate' and sand and rubble were pooled together into 'mixed sand' as they were not separated from one another in the 2000 survey. Additionally, transects from the 2000 study continued until the edge of the reef platform was reached, causing transects to consist of varying lengths (5–34 m) dependent on the width of the reef. The locations of transect sites (n = 60) during the 2000 survey were resurveyed in 2018 using a Garmin GPSMAP 78s; 3 m accuracy (Garmin Ltd., Olathe, KS, USA). Transects were spaced 10 meters apart to survey the 600-meter portion of the fringing reef (Figure 1). Both surveys were conducted with one snorkeling observer identifying all species in situ. Two community descriptors, cover and community composition, are used to empirically describe resilience to environmental stressors present at the site [14].

Figure 1. Map of Malauka'a fringing reef with transects overlaid within Kāne'ohe Bay, O'ahu. Note the variation in transect length due to reef width. Photo Credit: Digital Globe.

2.2.2. Seawater Temperature

Daily mean seawater temperatures (°C) for 2000 to 2018 in Kāne'ohe Bay were calculated from PacIOOS Moku o Lo'e weather station (http://www.pacioos.hawaii.edu/weather/obs-mokuoloe/).

2.2.3. Statistical Analysis

A two-tailed t-test was used to determine changes in daily average temperatures between 2000 and 2018 within RStudio IDE Version 1.1.456 (RStudio, Inc., Boston, MA, USA) [30]. A non-metric multidimensional scaling (NMDS) ordination plot using Bray–Curtis distance was created to visualize the 2000 and 2018 benthic communities within ggplot in RStudio [30]. A matched pair Wilcoxon signed-rank analysis was used to compare changes in individual species and groups (i.e., corals, algae, and mixed sand) between years (2000 vs. 2018) within transects using JMP Pro 13 (*JMP*®, Version 13, SAS Institute Inc., Cary, NC, USA) [31]. A permutational multivariate ANOVA (PERMANOVA) and a permutational test of multivariate dispersion (PERMDISP) were ran to determine if overall species composition changed between 2000 and 2018 using PERMANOVA+ [32] in PRIMER 7 Version 7.0.13 (PRIMER-e (Quest Research Limited) Auckland, New Zealand) [33]. The data for the PERMANOVA and PERMDISP was square root transformed before calculating a Bray–Curtis similarity matrix. The PERMANOVA was ran with two factors- fixed factor 'year' (2 levels, 999 unique permutations) and random 'transect' (6 levels, with transects pooled into 6 groups of 10 based on location, 998 unique permutations) nested in 'year.'

3. Results

3.1. Benthic Survey

Transects ranged from 5 to 34 meters in length, with 1219 observations recorded at one-meter intervals along the fringing reef in both 2000 and 2018. Six species of coral (i.e., *Porites compressa*, *Porites lobata*, *Montipora capitata*, *Lobactis* (formally *Fungia*) *scutaria*, *Pocillopora damicornis*, *Pocillopora meandrina*) were recorded at the site in 2000 and four (i.e., *P. compressa*, *M. capitata*, *P. damicornis*, *Leptastrea purpurea*) were recorded in 2018. Four species of macroalgae (i.e., *Dictyosphaeria cavernosa*, *Dictyosphaeria versluyii*, *Gracilaria salicornia*, *Kappaphycus alvarezii*) were present in 2000 and two (i.e., *D. cavernosa*, *D. versluyii*) were present in 2018. Unidentified species of turf algae, crustose coralline algae, and mixed sand and rubble were present in both surveys and were marked as such.

3.2. Statistical Analysis

3.2.1. Abiotic and Biotic Changes

The mean daily temperature (mean ± SE) at Moku o Lo'e increased from 24.07 ± 0.07 °C in 2000 to 25.03 ± 0.02 °C in 2018 ($p < 0.0001$), despite no evident general trend across years (R^2=0.1852) (Figure 2). The overall community composition across the fringing reef changed from 2000 to 2018 (PERMANOVA $p < 0.05$, PERMDISP $p < 0.05$) (Figure 3, Table 1). Total mixed sand cover decreased significantly from 12 ± 1.9% to 4.6 ± 1.0% from 2000 to 2018 ($p < 0.0001$) (Figures 4 and 5). This is further supported by a break in the fringing reef in 2000 (represented as a transect with 100% sand cover), which was not observed in the 2018 survey.

Table 1. PERMANOVA model results based on a Bray–Curtis similarity matrix comparing benthic communities between years (fixed factor) and transect section (random factor nested within year). Significant p values ($p < 0.05$) are bolded.

PERMANOVA							PERMDISP	
Source	df	SS	MS	Pseudo-F	*p*-Value	Unique Perms	F	*p*-Value
Year	1	25896	25896	5.3004	**0.003**	999	11.806	**0.004**
Transect (Year)	10	48975	4897.5	9.5632	**0.001**	998	9.8724	**0.001**
Residuals	108	55308	512.11					
Total	119	1.3004E+05						

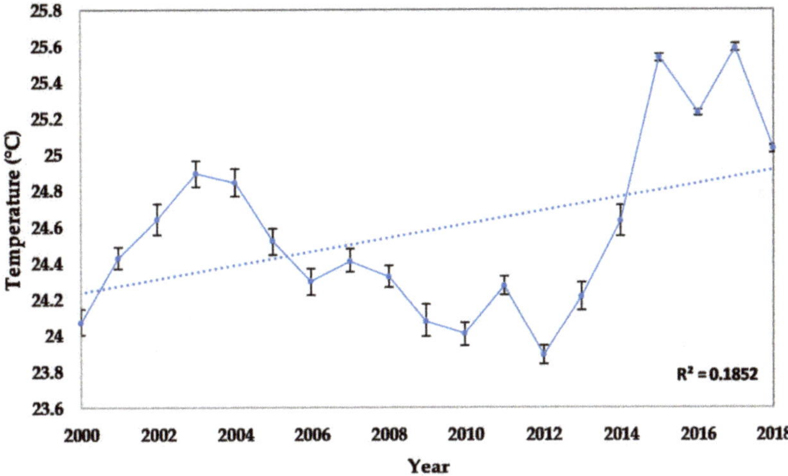

Figure 2. Average annual daily mean seawater temperature (°C) at Moku o Lo'e (Coconut Island) at the Hawai'i Institute of Marine Biology from 2000–2018. Data retrieved from (http://www.pacioos. hawaii.edu/weather/obs-mokuoloe/).

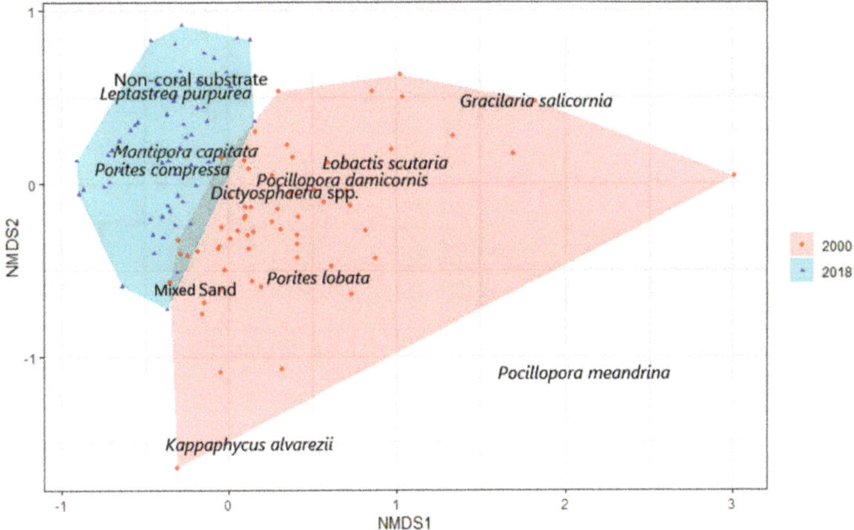

Figure 3. Non-metric multidimensional scaling (NMDS) ordination plot representing the benthic communities from the 2000 and 2018 surveys in convex hulls (Dimensions = 2, Stress = 0.19). Each point represents one transect (n = 60).

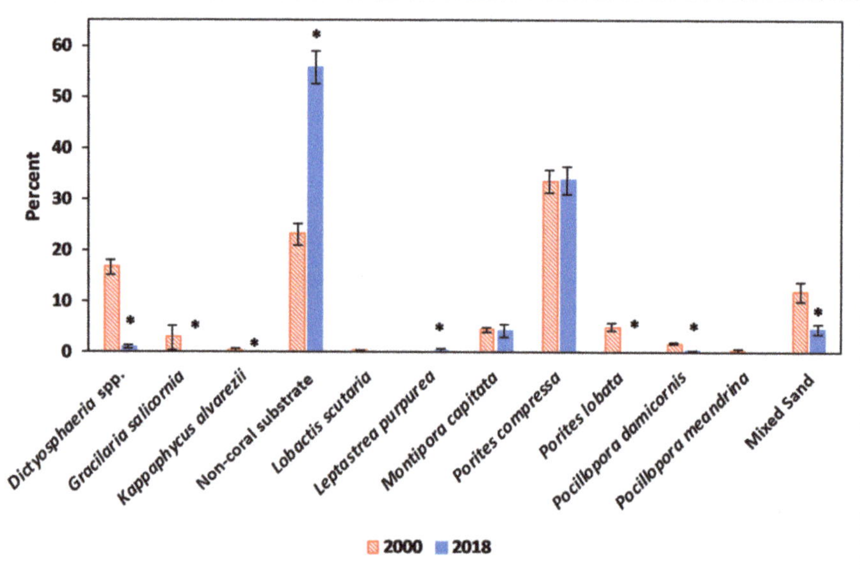

Figure 4. Mean Percent cover of each species or category in 2000 vs. 2018. Each standard error bar is one standard error from the mean. * Indicates significant difference between years at $p < 0.05$.

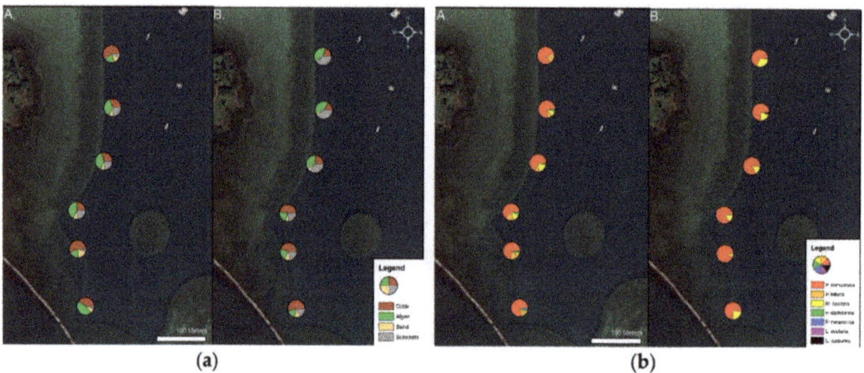

Figure 5. Spatial trends in (**a**) Total benthic cover and (**b**) coral species composition. (**A.**) represents data from the 2000 survey and (**B.**) represents data from the 2018 survey. Each pie chart represents the average from 10 transects in that section. Photo credit: Digital Globe

3.2.2. Algae

The total overall algae cover across the entire site increased significantly from 42.9 ± 3.1% in 2000 to 56.8 ± 3.2% in 2018 ($p = 0.0009$) (Figures 4 and 5). *Dictyosphaeria* spp. (*D. cavernosa* and *D. versluyii*) decreased significantly from 16.7 ± 1.5% in 2000 to 1.1 ± 0.3% 2018 ($p < 0.0001$). *Gracilaria salicornia* and *Kappaphycus alvarezii* were both present in 2000 (2.8 ± 2.4%, and 0.33 ± 0.3% respectively) and absent from the 2018 survey ($p = 0.0002$, 0.045). Non-coral substrate (turf, crustose coralline algae) increased significantly from 23.1 ± 2.1% in 2000 to 55.6 ± 3.2% in 2018 ($p < 0.0001$).

3.2.3. Coral

Total coral cover did not change between 2000 (45.1 ± 2.5%) and 2018 (38.6 ± 2.9%) (matched pair Wilcoxon signed-rank; $p = 0.0810$) (Figures 4 and 5). Neither dominant reef-building species (i.e., *Porites compressa* nor *Montipora capitata*) experienced a significant change in coverage percent. *Porites compressa* was found to cover 33.6 ± 2.3% and 33.7 ± 2.8% of the reef ($p = 0.8784$) and *M. capitata* was found to cover 4.4 ± 0.6% and 4.2 ± 1.2% ($p = 0.7836$) in 2000 and 2018, respectively. *Porites lobata* (5 ± 0.8%, $p < 0.0001$), *Pocillopora meandrina* (0.16 ± 0.4%, $p = 0.1590$), and *Lobactis scutaria* (0.16 ± 0.1%, $p = 0.1590$) were all present in the 2000 survey, but absent in 2018. *Lobactis scuatria* was visually observed at the site; however, it was not present on survey transects (personal observation, K.A.B., July 2018). *Pocillopora damicornis* decreased significantly from 1.8 ± 0.3% to 0.25 ± 0.1% from 2000 to 2018 ($p = 0.0005$). *Leptastrea purpurea* was not present in the 2000 survey but represented 0.49 ± 0.3% of the total cover in 2018 ($p = 0.0241$).

Spatial variations between 2000 and 2018 were also observed (Figure 5). The percent of coral cover was consistent between sections of the reef in 2000, whereas the percent of coral cover increased at the southern portion of the reef in 2018. In 2018, non-coral substrate was most common at the northern section of the reef, whereas it was more evenly distributed in 2000. *Montipora capitata* prevalence also increased in the southern portion of the reef from 2000 to 2018.

4. Discussion

While many reefs globally are in decline due to anthropogenic factors, coral cover on the reefs in Hawai'i remained stable from 1999–2012 [10]. Returning to the Malauka'a fringing reef provided an opportunity to explore decadal change in coral cover across an entire 600-meter reef. Results of this study revealed resilience and stability at the Malauka'a fringing reef over the past 18 years compared to other reefs across the Hawaiian islands. We predict the reef will show the same resilience as most reefs in Kāne'ohe Bay through maintaining high coral cover in the face of climate change.

4.1. Abiotic and Biotic Changes

During the 18 years between the two survey periods, corals at the study site experienced two consective bleaching events (i.e., 2014 and 2015). Seawater temperatures during these periods exceeded 31 °C for several days with cumulative heating of five degree heating weeks (DHW) in 2014 and 12 DHW in 2015 [11]. Between 2000 and 2018, daily average temperatures increased by 0.96 °C in Kāne'ohe Bay, indicating higher levels of temperature stress in 2018 compared to 2000.

The significant decrease in percent cover of mixed sand indicates the proportion of live benthic cover expanded between surveys.

4.2. Algae

Dictyosphaeria cavernosa was once the dominant algae species in Kāne'ohe Bay, responsible for one of the first well-studied reef phase shifts from coral-dominated to algae dominated [20]. Following the phase-shift reversal, the algae persisted in the bay due to overfishing of herbivorous fish that would have placed grazing pressure on the species [20]. *Dictyosphaeria cavernosa* remained abundant in Kāne'ohe Bay, averaging 16% total cover during a 1996–1997 survey [20]. The findings of the 2000 survey indicate the percent cover of *Dictyosphaeria* spp. remained at a comparable level three years later at the fringing reef (16.7 ± 1.5%). In 2006, following an unusually rainy period, decreased irradiance combined with slow spring growth rates for the species caused *D. cavernosa* to effectively disappear from Kāne'ohe Bay [34]. Immediately following the rapid decline, reefs nearby Moku o Lo'e averaged 0–4% total cover of *D. cavernosa* [35]. In 2018, twelve years later, the prevalence of *D. cavernosa* has remained greatly diminished at this fringing reef (1.1 ± 0.3%), suggesting an enduring phase shift reversal.

The invasive species *G. salicornia* was introduced to Kāne'ohe Bay in the 1970's and quickly spread, overgrowing and smothering reef-building corals [36]. The invasive algae has since decreased over the past few years as a result of biocontrol [37], manual removal [38], and increased grazing from *Chelonia mydas,* the green sea turtle [39]. The management efforts and return of *C. mydas* to Kāne'ohe Bay likely explain why the once dominant macroalgae was not observed during the 2018 survey.

Like *G. salicornia, Kappaphycus alvarezii* (formerly *Eucheuma striatum*) was introduced to Kāne'ohe Bay in the 1970's [40] and had spread across the southern and central bay by 1996 in a near-cosmopolitan distribution [41]. A total percent cover of 0.33 ± 0.3% in the 2000 survey was slightly higher than the mean 0.06 ± 0.02% cover found at four shallow fringing reefs in the central bay in 1996 [41]. Amidst fears of further spreading, preliminary management options for *Kappaphycus* spp. were assessed in 2002 [42]. Divers used an underwater vacuum cleaner and outplanted juvenile urchins (*Tripneustes gratilla*) to remove and control the species in 2011–2013, leading to an 85% decrease in invasive macroalgae across sites [38]. Management efforts have continued to be successful as *K. alvarezii* was not observed at the study site during the 2018 survey.

Despite *Dictyosphaeria* spp., *G. salicornia,* and *K. alvarezii* all decreasing or disappearing from the reef, a total increase in algal cover from 2000 to 2018 was observed, mainly due to the increase in 'non-coral substrate'. It should be noted that 18.6 ± 0.8% of the non-coral substrate from the 2018 survey was crustose coralline algae (CCA). CCA was not categorized or differentiated from 'encrusted corals' in the 2000 study. Thus, the percent cover of total algae as well as non-coral substrate is inflated in the 2018 data and likely the 2000 data as well. Unlike turf and macroalgae, CCA promotes coral recruitment and recovery [43] and would have ideally been separated into its own category.

The high percentage of non-coral substrate in 2018 (55.6 ± 0.9%) was also impacted by the prevalence of (perhaps short-lived) turf on the tips of *P. compressa* and *M. capitata.* The tips of these reef-building corals were susceptible to warming events and air exposure at extreme low tides as the 2018 survey was conducted in late July following a warm period and spring tides (Figure 6). Observed spatial differences within benthic communities showed certain sections of the reef were more susceptible to algal growth. During the 2018 survey, the northern portion of the reef exhibited higher levels of non-coral substrate than the southern portion of the reef (Figure 5). In addition to spatial variations in low tide air exposure, differences in temperatures could explain this occurrence as corals near the northern end experienced increased thermal stress (2018 summer midday average (11:00–16:00) temperature 27.72 ± 0.94) compared to corals at the southern end (2018 summer midday average temperature 27.48 ± 0.96). This difference highlights the importance local microclimates have on coral communities.

Figure 6. Reef Air Exposure. (**a**) Reef exposed during low tide in Kāne'ohe Bay (Picture Credit: KDB). (**b**) Tips of a pale *P. compressa* colony covered with turf (Picture Credit: KAB).

4.3. Corals

Despite a significant increase in algal cover between surveys, total coral cover was similar in 2000 and 2018. *Porites compressa* sustained a high percent cover over 18 years at the fringing reef despite decreasing in percent cover by 22.9% in 14 years (1999–2012) across the Hawaiian Islands, with significant declines on the island of Oʻahu [10]. *Porites compressa* is known to be sensitive to increased temperatures, which can cause bleaching and decreased calcification rates for the species [44]. Despite temperature increases over the 18 years, *P. compressa* has maintained its dominance as the most prevalent coral species at Malaukaʻa fringing reef, supporting its ability to acclimatize and persist in warming waters [24].

The *Montipora capitata* percent cover remained at a similar level between surveys despite increasing in percent cover by 56.8% in 14 years (1999–2012) across the Hawaiian Islands [10]. However, this study extended transects only to the end of the continuous reef pavement and many *M. capitata* colonies were located inshore of the reef (personal observation, K.A.B., July 2018). *Montipora capitata* colonies in Kāneʻohe Bay have shown resilience through the ability to acclimatize/adapt to temperature increases (2.6 °C) over the past 47 years [13]. The continued presence of *M. capitata* at Malaukaʻa fringing reef despite temperature increases supports the findings of [13] through indicating resilience in lab experimentation and field long-term monitoring.

Percent cover of *Pocillopora damicornis* decreased significantly between the 18 years. The species is known to be highly sensitive to decreased salinity levels [15]. Increased freshwater input onto the southern portion of the surveyed reef may have impacted the abundance of *P. damicornis*. Following biocultural restoration of the Paepae o Heʻeia, water exchange between the fishpond and the adjacent reef increased, with an additional 14,418 m^3 of pond water being flushed out onto the reef during each ebb tidal cycle [27].

In 2000, *P. lobata* was a common reef-building coral at the study site. However, *P. lobata* was not observed in the 2018 survey. *Porites lobata* was described as 'common to Kāneʻohe Bay' in 1999 [45]; however, more recently it was estimated to have 0–1% cover along Kāneʻohe's fringing reefs [46,47]. Previous work suggests that *P. lobata* and *P. compressa* are different morphotypes of the same species and/or hybridize frequently [48]. Therefore, the disappearance of *P. lobata* may mean one morphospecies was selected over the other. Due to similarities between *P. lobata* and *P. compressa* as well as the possibility of hybridizations, there may be potential misidentifications in the 2000 survey.

Similar to *P. lobata*, *P. meandrina* was also estimated to have 0–1% cover along fringing reefs in Kāneʻohe Bay, supporting its absence in the 2018 survey [46,47]. *Pocillopora meandrina* has been similarly decreasing in percent cover across the Hawaiian Islands, with a 36.1% decrease from 1999–2012 [10]. Following the 2015 bleaching event, 98% of *P. meandrina* colonies on the west side of the island of Hawaiʻi were partially or fully bleached, demonstrating they are one of the more susceptible species to thermal stress [49]. They were similarly listed as the least resistant species to thermal stress at Kahe Point, Oahu [50]. The species vulnerability to increased temperatures may explain its disappearance in the 2018 survey.

Lobactis (formely *Fungia*) *scutaria* was recorded during the 2000 survey but not observed in the 2018 survey. Low densities of *L. scutaria* are expected at the site, as the species is abundantly found on patch reefs in Kāneʻohe Bay, not fringing reefs [51]. Future studies of the area should employ a survey method such as the 'quadrat method', which avoids sampling from a small number of points to ensure rare and very rare species are included [28].

Leptastrea purpurea was the only new species seen in the 2018 survey. This encrusting species is tolerant to elevated temperatures and has been seen in areas where other coral species have succumbed to thermal stress [50]. The hardy species has been declared one of the 'long-term winners' as *L. purpurea* increase in abundance during thermal stress events [52,53]. *Leptastrea purpurea* has a relatively low metabolic rate, a characteristic known to help corals tolerate high temperatures [54]. Increasing temperatures may have allowed *L. purpurea* to settle in an area it had not before been present in, as it now holds a competitive advantage over other species which are less tolerant to thermal stress [53].

Coral cover did not significantly change over the past 18 years, although temperatures increased by 0.8 °C and two bleaching events (2014 and 2015) occurred during that time frame. While the fringing reef has shown resilience, it is unclear whether or not acclimatization and resistance to climate change has impacted its success. Previous work [13] has found all three species (i.e., *M. capitata*, *L. scutaria*, *P. damicornis*) of Hawaiian corals tested within Kāne'ohe Bay have higher survivorship at 31 °C today than they did in 1970, suggesting that these corals can adapt to higher temperatures. As the corals in this study were from similar locations as those used by References [13] and [24], it is possible the resilience seen on the reef can be attributed in part to adaptation or acclimatization. The persistence of the coral cover at this site occurred while other sites within Kāne'ohe Bay decreased in coral cover. From 2012–2016, Hawaii Coral Reef Assessment & Monitoring Program (CRAMP) reef sites at He'eia and Moku o Lo'e decreased by 19.7% and 42.2%, respectively [11].

However, while the total coral cover remained relatively stable over the past 18 years, the species composition has changed. The decrease in the total number of coral species present in the survey (6 in 2000, 4 in 2018) represents an overall loss in biodiversity. Additionally, two (or one if *P. lobata* is considered to be the same species as *P. compressa*) species of coral were lost in the 18 years while one non-reef building coral (*L. purpurea*) was added. This change suggests a temperature-driven shift in species composition over the 18 years. While the total coral cover remains high, the loss of locally uncommon species has negative impacts as rarer species often support more vulnerable and unique ecosystem functions [55].

Despite a shift in coral species composition, total coral cover percent remained unchanged over the 18 years and populations of the two dominant species of coral remained at comparable levels. Despite evidence of Hawaiian coral adaptation to increased temperatures, this adaptation might not occur fast enough to tolerate projected increasingly frequent bleaching events [13]. While the Malauka'a fringing reef has shown resilience over the past 18 years, the amount of warming and the rate of temperature increase will determine the fate of these reefs.

Author Contributions: Conceptualization, K.A.B. and K.D.B.; methodology, K.A.B. and K.D.B.; software, K.A.B. and K.D.B.; validation, K.A.B. and K.D.B.; formal analysis, K.A.B.; investigation, K.A.B.; resources, K.D.B.; data curation, K.A.B. and K.D.B.; writing—original draft preparation, K.A.B.; writing—review and editing, K.D.B.; visualization, K.A.B. and K.D.B.; supervision, K.D.B.

Funding: K.A.B acknowledges NMBU's Noragric field stipend for travel support to conduct research and NMBU's Publishing Fund for supporting the article processing charges. K.D.B. also acknowledges NSF#OA14-16889 for salary support during the writing and publication process.

Acknowledgments: The authors would like to thank Ku'ulei Rodgers for field advice and Ian Bryceson for his feedback on early drafts. Thanks to Annette Breckwoldt for providing data from the original 2000 survey. Thanks to Snorre Sundsbø and Elildo AR Carvalho Jr for assistance in statistical analyses. Thanks to Ashley McGowan, Colleen Brown, Robert Barnhill and members of the Coral Reef Ecology Lab for field and logistical support. Water temperature data was provided in part by PacIOOS (http://www.pacioos.org), which is a part of the U.S. Integrated Ocean Observing System (IOOS®), funded in part by National Oceanic and Atmospheric Administration (NOAA) Award ##NA16NOS0120024. We also thank three anonymous reviewers. We appreciate the time that each has invested to provide the review, and these comments have helped us improve the manuscript significantly.

Conflicts of Interest: The authors declare no conflict of interest.

References

1. Hoegh-Guldberg, O.; Mumby, P.J.; Hooten, A.J.; Steneck, R.S.; Greenfield, P.; Gomez, E.; Knowlton, N. Coral Reefs Under Rapid Climate Change and Ocean Acidification. *Science* **2007**, *318*, 1737–1742. [CrossRef] [PubMed]

2. Jokiel, P.L.; Coles, S.L. Response of Hawaiian and other Indo-Pacific Reef Corals to Elevated Temperature. *Coral. Reefs* **1990**, *8*, 155–162. [CrossRef]

3. Baker, A.C.; Glynn, P.W.; Riegl, B. Climate Change and Coral Reef Bleaching: An Ecological Assessment of Long-term Impacts, Recovery Trends and Future Outlook. *Estuar. Coast. Shelf Sci.* **2008**, *80*, 435–471. [CrossRef]

4. Graham, N.A.; Wilson, S.K.; Jennings, S.; Polunin, N.V.; Bijoux, J.P.; Robinson, J. Dynamic Fragility of Oceanic Coral Reef Ecosystems. *Proc. Natl. Acad. Sci. USA* **2006**, *103*, 8425–8429. [CrossRef] [PubMed]

5. Moberg, F.; Folke, C. Ecological Goods and Services of Coral Reef Ecosystems. *Ecol. Econ.* **1999**, *29*, 215–233. [CrossRef]

6. De'ath, G.; Fabricius, K.E.; Sweatman, H.; Puotinen, M. The 27–year Decline of Coral Cover on the Great Barrier Reef and its Causes. *Proc. Natl. Acad. Sci. USA* **2012**, *109*, 17995–17999. [CrossRef] [PubMed]

7. Wilkinson, C. *Status of Coral Reefs of the World: 2000*; Australian Institute of Marine Science: Queensland, Australia, 2000.

8. Bruno, J.F.; Selig, E.R. Regional Decline of Coral Cover in the Indo-Pacific: Timing, Extent, and Subregional Comparisons. *PLoS ONE* **2007**, *2*, e711. [CrossRef]

9. Burke, L.; Reytar, K.; Spalding, M.; Perry, A. *Reefs at Risk*; World Resources Institute: Washington, DC, USA, 2011; p. 124.

10. Rodgers, K.S.; Jokiel, P.L.; Brown, E.K.; Hau, S.; Sparks, R. Over a Decade of Change in Spatial and Temporal Dynamics of Hawaiian Coral Reef Communities. *Pac. Sci.* **2015**, *69*, 1–13. [CrossRef]

11. Bahr, K.D.; Rodgers, K.S.; Jokiel, P.L. Impact of Three Bleaching Events on the Reef Resiliency of Kāne'ohe Bay, Hawai'i. *Front. Mar. Sci.* **2017**, *4*. [CrossRef]

12. Hughes, T.P.; Kerry, J.T.; Álvarez-Noriega, M.; Álvarez-Romero, J.G.; Anderson, K.D.; Baird, A.H.; Bridge, T.C.; Butler, L.R.; Byrne, M.; Cantin, N.E.; et al. Global Warming and Recurrent Mass Bleaching of Corals. *Nature* **2017**, *543*, 373–377. [CrossRef]

13. Coles, S.L.; Bahr, K.D.; Ku'ulei, S.R.; May, S.L.; McGowan, A.E.; Tsang, A.; Han, J.H.; Bumgarner, J. Evidence of Acclimatization or Adaptation in Hawaiian Corals to Higher Ocean Temperatures. *PeerJ* **2018**, *6*. [CrossRef] [PubMed]

14. Bianchi, C.N.; Morri, C.; Lasagna, R.; Montefalcone, M.; Gatti, G.; Parravicini, V.; Rovere, A. Resilience of the Marine Animal Forest: Lessons from Maldivian Coral Reefs After the Mass Mortality of 1998. In *Marine Animal Forests: The Ecology of Benthic Biodiversity Hotspots*; Rossi, S., Bramanti, L., Gori, A., Orejas, C., Eds.; Springer: Cham, Switzerland, 2017; pp. 1241–1269. [CrossRef]

15. Jokiel, P.L.; Hunter, C.L.; Taguchi, S.; Watarai, L. Ecological Impact of a Fresh-water "Reef Kill" in Kaneohe Bay, Oahu, Hawaii. *Coral Reefs* **1993**, *12*, 177–184. [CrossRef]

16. Bahr, K.D.; Jokiel, P.L.; Toonen, R.J. The Unnatural History of Kāne'ohe Bay: Coral Reef Resilience in the Face of Centuries of Anthropogenic Impacts. *PeerJ* **2015**, *3*. [CrossRef] [PubMed]

17. Stimson, J. Recovery of Coral Cover in Records Spanning 44 Yr for Reefs in Kāne'ohe Bay, Oa'hu, Hawai'i. *Coral Reefs* **2018**, *37*, 55–69. [CrossRef]

18. Banner, A.H. Kaneohe Bay, Hawaii; Urban Pollution and a Coral Reefs Ecosystem. In Proceedings of the 2nd International Coral Reef Symposium, The Great Barrier Reef Province, Australia, 22 June–2 July 1974; Volume 2.

19. Smith, S.V.; Kimmerer, W.J.; Laws, E.A.; Brock, R.E.; Walsh, T.W. Kaneohe Bay Sewage Diversion Experiment: Perspective, on Ecosystem Responses to Nutritional Perturbation. *Pac. Sci.* **1981**, *35*, 279–395.

20. Stimson, J.; Larned, S.; Conklin, E. Effects of Herbivory, Nutrient Levels, and Introduced Algae on the Distribution and Abundance of the Invasive Macroalga Dictyosphaeria cavernosa in Kaneohe Bay, Hawaii. *Coral Reefs* **2001**, *19*, 343–357. [CrossRef]

21. Hunter, C.L.; Evans, C.W. Coral Reefs in Kaneohe Bay, Hawaii: Two Centuries of Western Influence and Two Decades of Data. *Bull. Mar. Sci.* **1995**, *57*, 501–515.

22. Jokiel, P.L.; Brown, E.K. Global Warming, Regional Trends and Inshore Environmental Conditions Influence Coral Bleaching in Hawaii. *Glob. Chang. Biol.* **2004**, *10*, 1627–1641. [CrossRef]

23. Bahr, K.D.; Jokiel, P.L.; Rodgers, K.S. The 2014 Coral Bleaching and Freshwater Flood Events in Kāne'ohe Bay, Hawai'i. *PeerJ* **2015**, *3*. [CrossRef]

24. Jury, C.P.; Toonen, R.J. Adaptive Responses and Local Stressor Mitigation Drive Coral Resilience in Warmer, More Acidic Oceans. *Proc. R. Soc. Lond. B. Biol. Sci.* **2019**, *286*. [CrossRef]

25. Mühlig-Hofmann, A. The Distribution of Corals and Macroalgae in Relation to Freshwater and Sediment Discharge on a Fringing Reef in Kane'ohe Bay, O'ahu, Hawai'i. Master's Thesis, University of Bremen, Bremen, Germany, 2001.

26. Bremer, L.; Falinski, K.; Ching, C.; Wada, C.; Burnett, K.; Kukea-Shultz, K.; Reppun, N.; Chun, G.; Oleson, K.L.L.; Ticktin, T. Biocultural Restoration of Traditional Agriculture: Cultural, Environmental, and Economic Outcomes of Lo 'i Kalo Restoration in He'eia, O 'ahu. *Sustainability* **2018**, *10*, 4502. [CrossRef]

27. Möhlenkamp, P.; Beebe, C.; McManus, M.; Kawelo, A.; Kotubetey, K.; Lopez-Guzman, M.; Nelson, C.E.; Alegado, R. Kū Hou Kuapā: Cultural Restoration Improves Water Budget and Water Quality Dynamics in He'eia Fishpond. *Sustainability* **2019**, *11*, 161. [CrossRef]

28. Hill, J.; Wilkinson, C. *Methods for Ecological Monitoring of Coral Reefs*; Australian Institute of Marine Science: Queensland, Australia, 2004; Volume 117, ISBN 0-642-32237-6.

29. Jokiel, P.L.; Rodgers, K.S.; Brown, E.K.; Kenyon, J.C.; Aeby, G.; Smith, W.R.; Farrell, F. Comparison of Methods Used to Estimate Coral Cover in the Hawaiian Islands. *PeerJ* **2015**, *3*. [CrossRef] [PubMed]

30. RStudio Team. *RStudio: Integrated Development for R*; RStudio, Inc.: Boston, MA, USA, 2015; Available online: http://www.rstudio.com/ (accessed on 1 May 2019).

31. *JMP*®, version 13; SAS Institute Inc.: Cary, NC, USA, 1989–2019.

32. Anderson, M.J.; Gorley, R.N.; Clarke, K.R. *PERMANOVA + for PRIMER: Guide to Software and Statistical Methods*; PRIMER-E: Plymouth, UK, 2008.

33. Clarke, K.R.; Gorley, R.N. *PRIMER v7: User Manual/Tutorial*; PRIMER-E: Plymouth, UK, 2015.

34. Jokiel, P.L.; Coles, S.L. Effects of Heated Effluent on Hermatypic Corals at Kahe Point, Oahu. *Pac. Sci.* **1974**, *28*, 1–18.

35. Stimson, J.; Conklin, E.J. Potential Reversal of a Phase Shift: The Rapid Decrease in the Cover of the Invasive Green Macroalga Dictyosphaeria cavernosa Forsskål on Coral Reefs in Kāne 'ohe Bay, Oahu, Hawai 'i. *Coral Reefs* **2008**, *27*, 717–726. [CrossRef]

36. Smith, J.E.; Hunter, C.L.; Conklin, E.J.; Most, R.; Sauvage, T.; Squair, C.; Smith, C.M. Ecology of the Invasive Red Alga Gracilaria salicornia (Rhodophyta) on O'ahu, Hawai'i. *Pac. Sci.* **2004**, *58*, 325–343. [CrossRef]

37. Stimson, J.; Cunha, T.; Philippoff, J. Food Preferences and Related Behavior of the Browsing Sea Urchin Tripneustes gratilla (Linnaeus) and its Potential for use as a Biological Control Agent. *Mar. Biol.* **2007**, *151*, 1761–1772. [CrossRef]

38. Neilson, B.J.; Wall, C.B.; Mancini, F.T.; Gewecke, C.A. Herbivore Biocontrol and Manual Removal Successfully Reduce Invasive Macroalgae on Coral Reefs. *PeerJ* **2018**, *6*. [CrossRef] [PubMed]

39. Bahr, K.D.; Coffey, D.M.; Rodgers, K.S.; Balazs, G.H. Observations of a Rapid Decline in Invasive Macroalgal Cover Linked to Green Turtle Grazing in a Hawaiian Marine Reserve. *Micronesica* **2018**, *7*, 1–11.

40. Russell, D.J. Ecology of the Imported Red Seaweed Eucheuma striatum Schmitz on Coconut Island, Oahu, Hawaii. *Pac. Sci.* **1983**, *37*, 87–107.

41. Rodgers, K.S.; Cox, E.F. Rate of Spread of Introduced Rhodophytes Kappaphycus alvarezii, Kappaphycus striatum, and Gracilaria salicornia and Their Current Distribution in Kane'ohe Bay, O'ahu Hawai'i. *Pac. Sci.* **1999**, *53*, 232–241.

42. Conklin, E.J.; Smith, J.E. Abundance and Spread of the Invasive Red Algae, Kappaphycus spp., in Kane'ohe Bay, Hawai'i and an Experimental Assessment of Management Options. *Biol. Invasions* **2005**, *7*, 1029–1039. [CrossRef]

43. Price, N. Habitat Selection, Facilitation, and Biotic Settlement Cues Affect Distribution and Performance of Coral Recruits in French Polynesia. *Oecologia* **2010**, *163*, 747–758. [CrossRef]

44. Carricart-Ganivet, J.P.; Cabanillas-Teran, N.; Cruz-Ortega, I.; Blanchon, P. Sensitivity of Calcification to Thermal Stress Varies Among Genera of Massive Reef-building Corals. *PLoS ONE* **2012**, *7*, e32859. [CrossRef] [PubMed]

45. Grottoli, A.G. Variability of Stable Isotopes and Maximum Linear Extension in Reef-coral Skeletons at Kaneohe Bay, Hawaii. *Mar. Biol.* **1999**, *135*, 437–449. [CrossRef]

46. Franklin, E.C.; Jokiel, P.L.; Donahue, M.J. Predictive Modeling of Coral Distribution and Abundance in the Hawaiian Islands. *Mar. Ecol. Prog. Ser.* **2013**, *481*, 121–132. [CrossRef]

47. Franklin, E.C.; Jokiel, P.L.; Donahue, M.J. Data from: Predictive Modeling of Coral Distribution and Abundance in the Hawaiian Islands. *Dryad Digit. Repos.* **2013**. [CrossRef]

48. Forsman, Z.H.; Knapp, I.S.S.; Tisthammer, K.; Eaton, D.A.R.; Belcaid, M.; Toonen, R.J. Coral Hybridization or Phenotypic Variation? Genomic Data Reveal Gene Flow Between Porites lobata and P. Compressa. *Mol. Phylogenet. Evol.* **2017**, *111*, 132–148. [CrossRef] [PubMed]

49. Maynard, J.; Conklin, E.J.; Minton, D.; Most, R.; Couch, C.S.; Williams, G.J.; Gove, J.M.; Dieter, T.; Schumacher, B.; Martinez, J.; et al. *Relative Resilience Potential and Bleaching Severity in the West Hawai'i Habitat Focus Area in 2015*; National Oceanic and Atmospheric Administration: Washington, DC, USA, 2016. [CrossRef]
50. Jokiel, P.L.; Maragos, J.E. Reef Corals of Canton Atoll: II. Local distribution. In *Phoenix Islands Report I. An. Environmental Survey of Canyon Atoll Lagoon 1973*; Smith, S.V., Henderson, R.S., Eds.; Smithsonian Institute: Washington, DC, USA, 1978; Volume 221, pp. 73–97.
51. Lacks, A. Reproductive Ecology and Distribution of the Scleractinian coral Fungia Scutaria in Kane'ohe bay, O'ahu, Hawai'i. Master's Thesis, University of Hawaii, Manoa, HI, USA, 2000.
52. Van Woesik, R.; Sakai, K.; Ganase, A.; Loya, Y. Revisiting the Winners and the Losers a Decade after Coral Bleaching. *Mar. Ecol. Prog Ser.* **2011**, *434*, 67–76. [CrossRef]
53. Bahr, K.D.; Rodgers, K.S.; Jokiel, P.L. Relative Sensitivity of Five Hawaiian Coral Species to High Temperature Under High-pCO2 Conditions. *Coral Reefs* **2016**, *35*, 729–738. [CrossRef]
54. Mayer, A.G. Is Death from High Temperature Due to the Accumulation of Acid in the Tissues? *Am. J. Physiol. Leg. Content* **1917**, *44*, 581–585. [CrossRef]
55. Mouillot, D.; Bellwood, D.R.; Baraloto, C.; Chave, J.; Galzin, R.; Harmelin-Vivien, M.; Kulbicki, M.; Lavergne, S.; Lavorel, S.; Paine, C.T.; et al. Rare Species Support Vulnerable Functions in High-diversity Ecosystems. *PLoS Biol.* **2013**, *11*, e1001569. [CrossRef] [PubMed]

Journal of
*Marine Science
and Engineering*

Article

Assessing the Resilience Potential of Inshore and Offshore Coral Communities in the Western Gulf of Thailand

Makamas Sutthacheep, Charernmee Chamchoy, Sittiporn Pengsakun, Wanlaya Klinthong and Thamasak Yeemin *

Marine Biodiversity Research Group, Department of Biology, Faculty of Science, Ramkhamhaeng University, Huamark, Bangkok 10240, Thailand; smakamas@hotmail.com (M.S.); charernmee14@hotmail.com (C.C.); marine_ru@hotmail.com (S.P.); klinthong_fai@hotmail.com (W.K.)
* Correspondence: thamasakyeemin@hotmail.com; Tel.: +6623108415

Received: 21 July 2019; Accepted: 7 November 2019; Published: 11 November 2019

Abstract: Coral reefs in the Gulf of Thailand have experienced severe coral bleaching events and anthropogenic disturbances during the last two decades. This study assessed the resilience potential of coral communities at Ko Losin offshore reef sites and Mu Ko Chumphon nearshore coral reefs, in the south of Thailand, by conducting field surveys on the live coral cover, hard substratum composition and diversity and density of juvenile corals. Most study sites had higher percentages of live coral cover compared to dead coral cover. Some inshore and offshore reef sites showed low resilience to coral bleaching events. The total densities of juvenile corals at the study sites were in the range of 0.89–3.73 colonies/m^2. The density of the juvenile corals at most reef sites was not dependent on the live coral cover of adult colonies in a reef, particularly for the *Acropora* communities. We suggest that Ko Losin should be established as a marine protected area, and Mu Ko Chumphon National Park should implement its management plans properly to enhance coral recovery and promote marine ecotourism. Other measures, such as shading, should be also applied at some coral reefs during bleaching periods.

Keywords: coral; recruitment; resilience; bleaching; management; restoration; fishing; tourism; recovery; Thailand

---

1. Introduction

Coral reefs are recognized as a high-biodiversity ecosystem containing thousands of species that provide socioeconomic benefits. The benefits include providing food and livelihoods for millions of people in tropical countries and the protection of coastal communities from extreme weather disturbances [1,2]. However, coral reefs around the world are degrading because of natural stressors (bleaching, diseases and heavy storms [3–9]) and anthropogenic disturbances, particularly coastal development, pollution, sedimentation and overfishing [10–13]. Human impacts have also reduced the ability of coral recovery and reef resilience after severe disturbances [14–16]. Knowledge about the synergistic effects of coral bleaching and human activities on the ecological processes of coral reefs, particularly coral recruitment, is very important for establishing a science-based management strategy for enhancing the resilience potential of coral reefs [17].

Coral reef management requires supporting ecosystem processes that lower sensitivity, promote recovery, and enhance the adaptive capacity of coral reefs to bleaching by reducing other human impacts [18]. The capacity of coral reefs to resist or recover from degradation and to maintain their ecosystem services is defined as coral reef resilience [19]. Resilience-based management of coral reefs includes assessing spatial variation in resilience potential and implementing appropriate management plans [18,20,21]. The assessment of the resilience potential of coral reefs was first developed after the coral bleaching event in the year 1998 and it focused on the physical and ecological characteristics of coral reefs that provide some reefs with greater resistance to and/or recovery from coral bleaching [22,23]. Several resilience indicators have been widely developed and proposed for assessing the ecological resilience of coral reefs [24–27].

Successful coral recruitment and juvenile survivorship play an important role in the maintenance of coral populations under normal natural conditions and following mass mortality from bleaching events [28–30]. The planktonic larval stage, settlement and juvenile coral are critical periods in the coral life cycle and have high mortality rates, particularly under stressful environments. Following coral bleaching events, most surviving adult corals show reduced fecundity and growth as well as decreased reproductive outputs and recruitment rates [18,31]. Therefore, coral recruitment is often used as a bioindicator of coral reef health, recovery rate and resilience potential after severe disturbances such as bleaching events. A high coral recruitment rate or high density of juvenile corals on natural substrates can lead to quick coral recovery of degraded reefs after coral bleaching events and anthropogenic disturbances [32]. Coral recovery is also controlled through grazing by herbivores, which limits algal growth [33]. Several environmental factors influence coral recruitment rates, particularly water pollution, overfishing and coastal development, which can affect coral competition ability, fecundity, fertilization success, settlement and survival of juvenile corals [34–36]. Coral recovery and the resilience potential of coral reefs are usually controlled by coral larval supply, recruitment rate, the survival rate of juvenile corals and high resistance/tolerance to environmental stresses [17,37,38].

Mass coral bleaching events in the Gulf of Thailand were reported in 1998, 2010 and 2016 [39–41]. There were significant differences in the susceptibility of coral species to bleaching events in the Gulf of Thailand between the years 1998 and 2010. The 2010 coral bleaching phenomenon at some reef sites, such as Ko Samui in the Western Gulf of Thailand, was more severe than the 1998 bleaching event [39]. The intensive study of coral bleaching in the Gulf of Thailand in the year 2016 revealed that the levels of coral bleaching varied significantly among the reef sites. A high severity level of coral bleaching, of about 70%, was recorded at Ko Ngam Noi, Chumphon Province, in the south of Thailand. The coral mortality following the 2016 bleaching event was approximately 18%, which was much lower than that of the 2010 coral bleaching event because the southwest monsoon started earlier, and therefore the seawater temperature dropped rapidly [41]. Previous studies defined resilience as the capacity of a system to absorb or withstand stressors, maintain its structure and functions in the face of disturbance and change and adapt to future challenges [42,43]. This study aims to assess the resilience potential, based on coverages of live coral, dead coral, rubble and other benthic organisms, of coral communities at Ko Losin offshore reef sites in Pattani Province and Mu Ko Chumphon nearshore coral reefs in Chumphon Province, in the south of Thailand. Field surveys on the live coral cover, hard substratum composition and diversity and density of juvenile corals were conducted to determine the resilience of the coral communities in the south of Thailand.

2. Materials and Methods

The study was conducted on coral communities in the Western Gulf of Thailand in March–May 2019. Six study sites from two different groups of coral communities, i.e., three study sites from Ko Losin offshore coral assemblages on pinnacles and three study sites from Mu Ko Chumphon nearshore coral reefs in Mu Ko Chumphon National Park, were selected for this study (Figure 1). Ko Losin is a small isolated island with an old lighthouse giving signals to boat navigators, about 72 km from the mainland. It has a relatively high water clarity in the Gulf of Thailand and harbors coral reefs that are well developed in deeper water, extending from 7 to 25 m depth. Ko Losin has been affected by fishing activities as it is an unprotected remote area. Recently, it is also used as a diving site in the Gulf of Thailand during the southwest monsoon period. Mu Ko Chumphon National Park is a marine protected area that is managed by the Department of National Parks, Plant and Wildlife Conservation. There are about 40 nearshore islands in Chumphon Province in the Western Gulf of Thailand, which harbor several coral reefs in good condition with high potential for tourism, particularly snorkeling and SCUBA diving. Three reef sites in Mu Ko Chumphon, i.e., Ko Kula, Ko Ngam Yai and Ko Ngam Noi, were selected for the field surveys. The coral reefs at the study sites were in shallow water, 1–12 m in depth. Ko Kula had relatively turbid water as it was affected by high sediment load from the mainland. The location, environmental conditions and anthropogenic disturbances at each study site are summarized in Table 1.

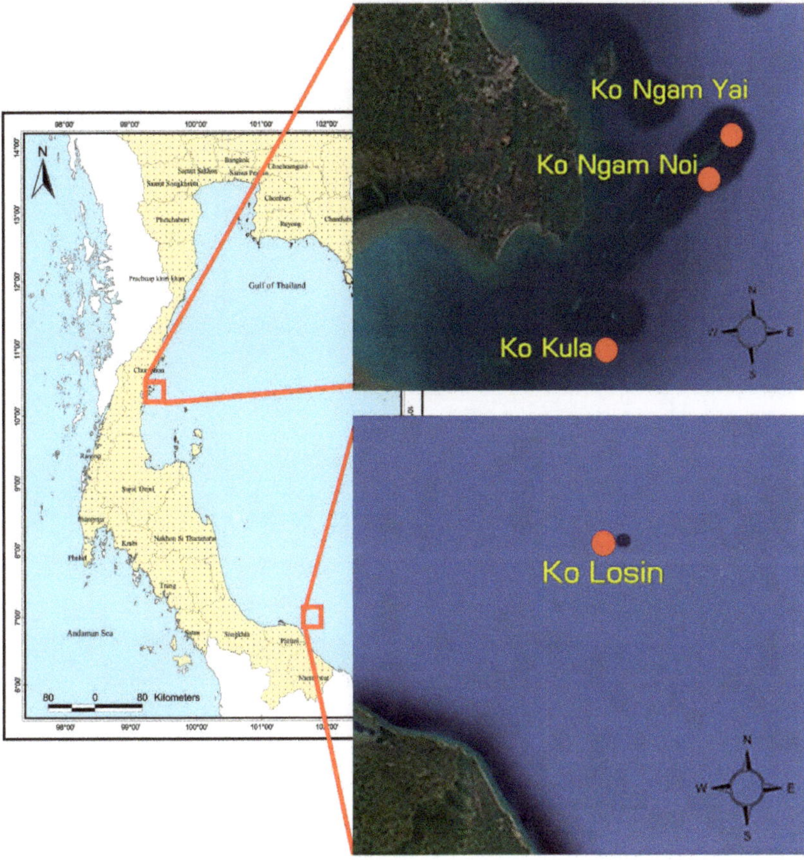

Figure 1. Map of the study sites at Mu Ko Chumphon National Park, Chumphon Province and Ko Losin, Pattani Province.

Table 1. Location and information of the study sites in the south of Thailand.

Study Sites	Latitude (N), Longitude (E)	Exposure Condition	Coral Reef Type	Distance from the Shore (km)	Water Transparency	Depth (m)	Anthropogenic Disturbances
Pattani Province							
Ko Losin (West)	7°19.376′ N 101°53.298′ E	Exposed	Developing reef	72	Clear	8–25	Tourism (low), Fishery (high)
Ko Losin (South)	7°18.830′ N 101°53.900′ E	Exposed	Developing reef	72	Clear	10–20	Tourism (low), Fishery (high)
Ko Losin (East)	7°19.484′ N 101°54.340′ E	Exposed	Developing reef	72	Clear	7–20	Tourism (low), Fishery (high)
Chumphon Province							
Ko Kula	10°15.347′ N 99°15.205′ E	Sheltered	Fringing reef	5.5	Turbid	1–5	Tourism (high), Fishery (low)
Ko Ngam Yai	10°29.531′ N 99°25.120′ E	Sheltered	Fringing reef	21	Clear	1–6	Tourism (high), Fishery (low)
Ko Ngam Noi	10°29.200′ N 99°25.060′ E	Sheltered	Fringing reef	20.5	Clear	1–12	Tourism (high), Fishery (low)

At each study site, the live coral cover was observed and evaluated as colony area/unit area in three belt-transects of 50×1 m^2, coral colonies (≥ 5 cm in diameter) were counted and identified to the species level [44], if possible, and their coverage was quantitatively estimated. Covers of dead corals, rubble, sand, rock and other benthic components were recorded. In this study, covers of dead corals, rubble, rock and other benthic components were combined as available substrate. Quadrats were also photographed with an underwater camera for reinvestigating the data. Quadrats (50×50 cm^2 each) were randomly placed on available substrates at each study site by SCUBA divers, and the number of juvenile coral colonies (≤ 5 cm in diameter) was carefully observed, identified, counted and photographed for reconfirmation in the laboratory. All juvenile coral colonies were identified to the genus level [44].

Cluster analysis and the non-multidimensional scaling method were performed to categorize study sites on the basis of the Bray–Curtis similarity of benthic components, using PRIMER version 7.0. Differences in the taxonomic composition of corals between Ko Losin and Mu Ko Chumphon were tested by analysis of similarities (ANOSIM), and the coral species contributing most to the dissimilarity between the study sites were identified by similarity percentage (SIMPER) analyses. A one-way ANOVA was used to test the differences in the percentages of live coral cover, species diversity and juvenile coral densities among study sites. Where significant differences were found, the Tukey HSD (honestly significant difference) test was employed to determine which reef site(s) differed.

3. Results

There were significant differences in coral cover among study sites (one-way ANOVA, $p < 0.05$) (Figures 2 and 3). The highest percentages of live coral cover were found at Ko Ngam Noi (77.3 ± 9.3) and Ko Kula (57.7 ± 6.9) in Mu Ko Chumphon and at Ko Losin (West) (47.0 ± 18.0), Ko Losin (East) (45.7 ± 20.5) and Ko Losin (South) (26.7 ± 10.2), while the lowest coverage was observed at Ko Ngam Yai (5.4 ± 0.6). All study sites except Ko Ngam Yai had a higher percentage of live coral cover compared to dead coral cover.

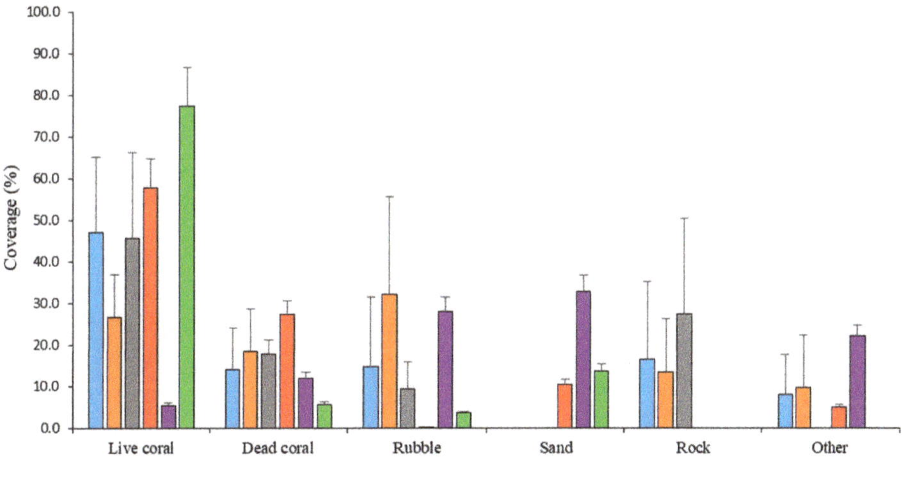

Figure 2. Average percentage cover of live corals, dead corals and other benthic components at the study sites. Error bars indicate standard deviation.

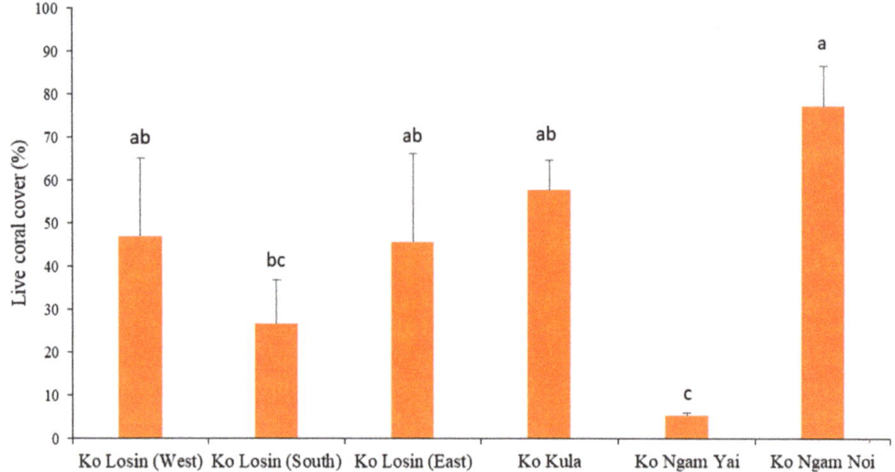

Figure 3. Live coral cover at the study sites (one-way ANOVA, $p < 0.05$). Error bars indicate standard deviation. Different letters above bars indicate statistical differences ($p < 0.05$), as determined by Tukey's HSD.

All reef sites except Ko Kula harbored relatively high coral diversity. The highest resilience potential site was Ko Ngam Noi, which was dominated by *Acropora* spp. The high potential sites included Ko Kula, Ko Losin (West) and Ko Losin (East), while the low resilience-potential sites were Ko Ngam Yai and Ko Losin (South), which were dominated by *Porites lutea* (Figure 4). Overall, only Ko Ngam Yai had low resilience potential in terms of survival after bleaching and anthropogenic disturbances. The Shannon–Wiener index of diversity (H′) was significantly different among the six study sites (one-way ANOVA, F = 25.27, $p = 0.001$). Tukey HSD tests showed that Ko Losin (East) was more diverse (H′ = 1.7 ± 0.2) than Ko Kula (H′ = 0.5 ± 0.1) (Figure 5).

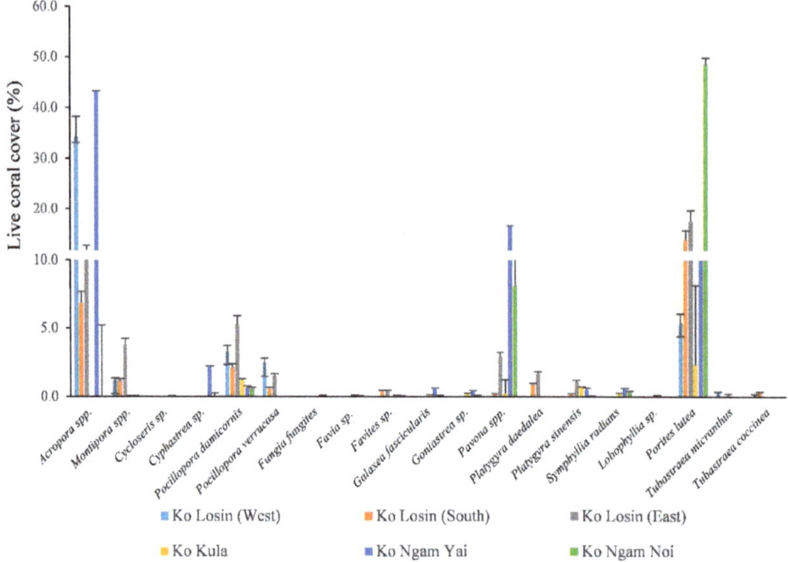

Figure 4. Species composition of corals at the study sites. Error bars indicate standard deviation.

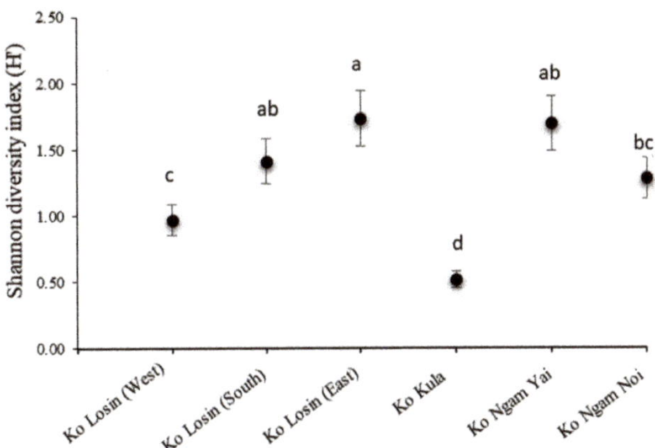

Figure 5. Shannon–Wiener index of diversity (mean ± SD) of coral species for each study site (one-way ANOVA, $p < 0.05$). Different letters above indicate statistical differences ($p < 0.05$), as determined by Tukey's HSD.

ANOSIM indicated significant differences in the taxonomic composition of corals between Ko Losin and Mu Ko Chumphon (R = 0.52, $p < 0.001$, Figure 6). The average similarity in the composition of coral species between Ko Losin and Mu Ko Chumphon ranged from about 41.64% to 69.62%, whereas dissimilarity between Ko Losin and Mu Ko Chumphon was 59.74% (SIMPER analysis), Table 2.

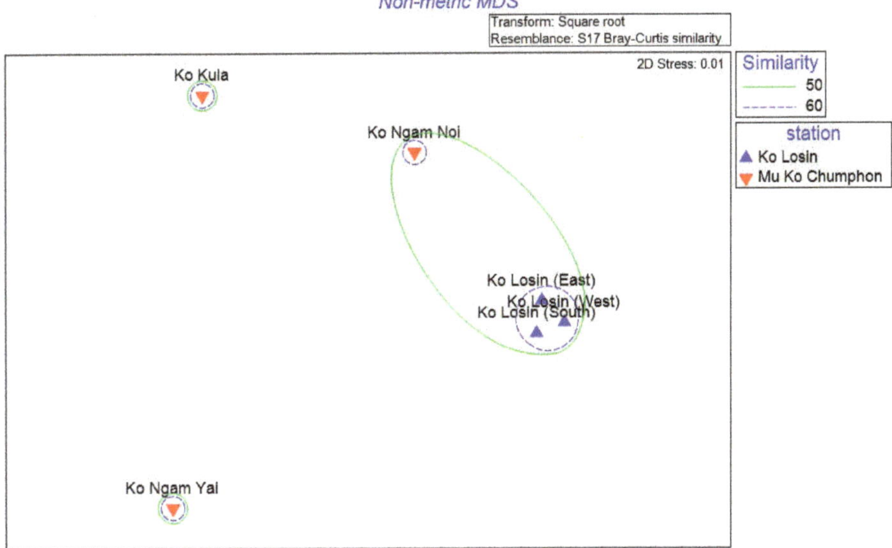

Figure 6. Two-dimensional non-metric multidimensional scaling (NMDS) plot of the taxonomic composition of corals at the study sites.

Table 2. Similarity percentage (SIMPER) analysis of benthic communities in two regions in the Gulf of Thailand.

SIMPER	Average Dissimilarity (%)
Ko Losin and Mu Ko Chumphon	
Acropora spp.	13.69
Porites lutea	8.16
Pavona spp.	6.71
Montipora spp.	4.54
Pocillopora verrucosa	4.43
Pocillopora damicornis	3.46
Platygyra daedalea	2.68
Symphyllia radians	2.22
Platygyra sinensis	1.80
Goniastrea sp.	1.59
Favites sp.	1.45
Cyphastrea sp.	1.36
Tubastraea coccinea	1.35
Galaxea fascicularis	1.31

The two-dimensional non-metric multidimensional scaling (NMDS) plot of the study sites based on the live corals, dead corals and other benthic components revealed that there were three groups of study sites, i.e., all three study sites of Ko Losin, Ko Kula and Ko Ngam Noi study sites, and Ko Ngam Yai study site (Figure 7).

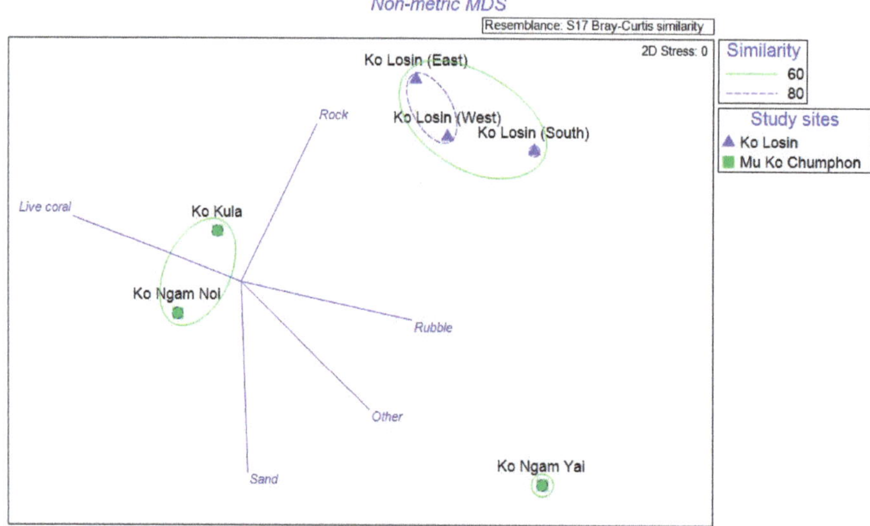

Figure 7. Two-dimensional NMDS plot of the study sites.

Underwater photographs of the six study sites are shown in Figure 8. All study sites at Ko Losin and Ko Ngam Noi still displayed high live coral cover of *Acropora* spp., indicating that these reef sites were highly resilient to the coral bleaching events in 1998, 2010 and 2016.

Figure 8. Underwater photographs showing the dominant coral species at the study sites.

The total densities of juvenile corals, i.e., those less than 5 cm in diameter, at the study sites were in the range of 0.89–3.73 colonies/m^2. The highest average density of juvenile corals was found at Ko Ngam Yai (3.73 colonies/m^2), while the lowest average density was found at Ko Losin (West) (0.89 colonies/m^2). The total density of juvenile corals at Ko Ngam Yai was significantly higher than that at Ko Ngam Noi, Ko Kula and all study sites of Ko Losin (one-way ANOVA; Tukey's HSD test; $p < 0.05$) (Figure 9). A total of 19 genera of juvenile corals were commonly observed, namely, *Pocillopora*, *Tubastrea*, *Montipora*, *Galaxea*, *Pavona*, *Pachyseris*, *Fungia*, *Lithophyllon*, *Hydnophora*, *Turbinaria*, *Lobophyllia*, *Favia*, *Favites*, *Oulastrea*, *Leptastrea*, *Cyphastrea*, *Porites*, *Goniopora* and *Plerogyra*. The juvenile corals of *Pocillopora* were dominant at all study sites except Ko Kula. The most dominant juvenile corals at the study sites of Ko Losin were *Pocillopora*, *Porites* and *Tubastrea*, while the dominant juvenile corals at the study sites of Mu Ko Chumphon were *Pocillopora*, *Porites*, *Fungia*, *Pachyseris*, *Pavona*, *Favites* and *Leptastrea* (Figure 10).

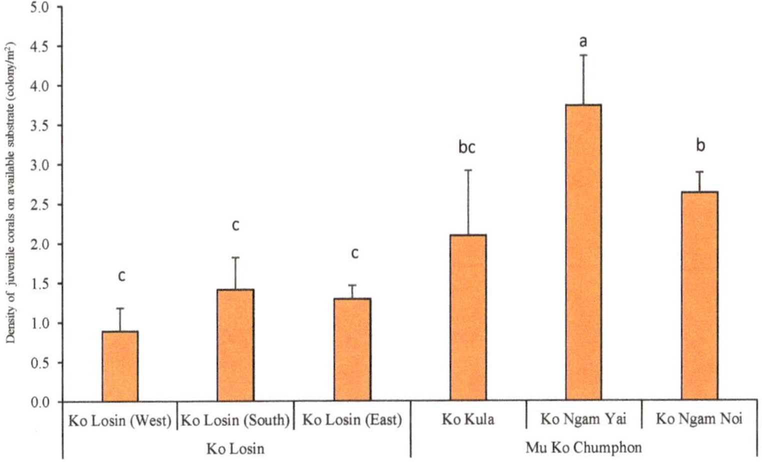

Figure 9. Densities of juvenile corals (mean ± SD) on available substrate at the study sites (one-way ANOVA, $p < 0.05$). Different letters above bars indicate statistical differences ($p < 0.05$), as determined by Tukey's HSD.

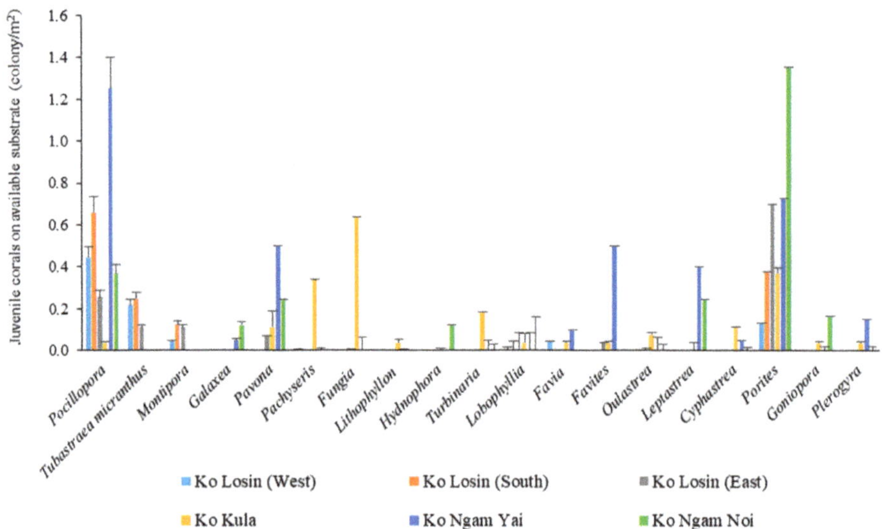

Figure 10. Composition of the juvenile corals on available substrate at the study sites. Error bars indicate standard deviation.

ANOSIM indicated significant differences in the composition of juvenile corals between Ko Losin and Mu Ko Chumphon (R = 0.63, *p* < 0.001, Figure 11). The average similarity in the composition of juvenile corals between Ko Losin and Mu Ko Chumphon ranged from about 43.17% to 73.68%, whereas dissimilarity between Ko Losin and Mu Ko Chumphon was 63.81% (SIMPER analysis), Table 3.

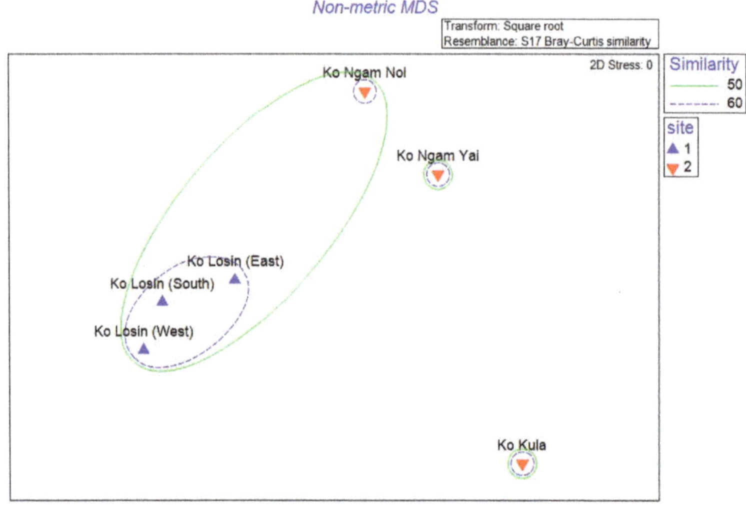

Figure 11. Two-dimensional NMDS plot of the composition of juvenile corals at the study sites.

Table 3. SIMPER analysis of the composition of juvenile corals at the study sites.

SIMPER	Average Dissimilarity (%)
Ko Losin and Mu Ko Chumphon	
Tubastraea micranthus	4.57
Pavona spp.	2.06
Leptastrea spp.	1.33
Porites spp.	1.15
Pocillopora spp.	1.79
Montipora spp.	4.60
Favites spp.	0.98
Fungia spp.	0.67
Goniopora spp.	1.09
Galaxea spp.	1.17
Pachyseris spp.	0.67
Plerogyra spp.	1.18
Cyphastrea spp.	1.24
Favia spp.	1.22
Turbinaria spp.	0.67

The juvenile coral densities of the brooder *Pocillopora* were relatively high at Ko Ngam Noi (0.37 ± 0.15 colonies/m^2), Ko Losin Pinnacle (South) (0.66 ± 0.08 colonies/m^2) and Ko Losin (West) (0.44 ± 0.05 colonies/m^2). The juvenile coral densities of broadcast spawners at the study sites of Mu Ko Chumphon were much higher compared to those at the study sites of Ko Losin (Figure 12). Underwater photographs of the dominant juvenile corals, *Pocillopora*, *Porites* and *Tubastraea*, at the six study sites are shown in Figure 13. The juvenile corals were in healthy conditions without any signs of partial mortality or stress from competitors, diseases and bleaching.

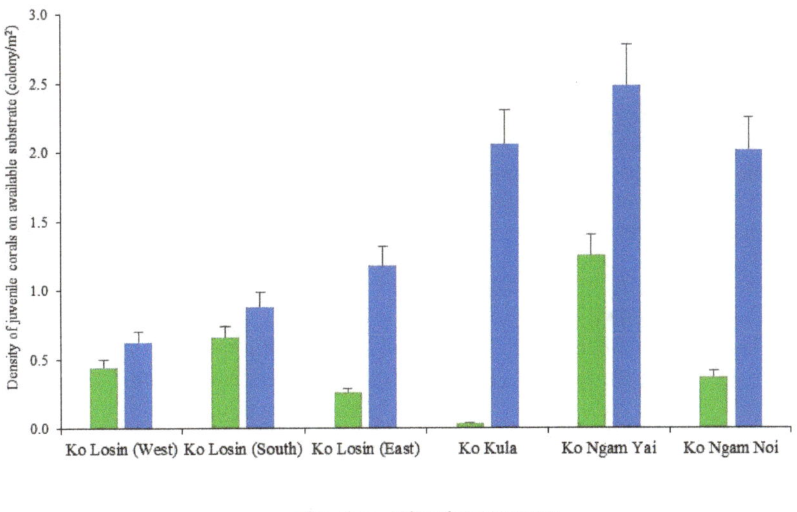

Figure 12. Densities of juvenile corals on available substrate for broadcast spawners and brooders at the study sites. Error bars indicate standard deviation.

Figure 13. Dominant juvenile corals on available substrate at the study sites.

4. Discussion

The coral reefs in the Gulf of Thailand are developed in high turbidity and have experienced severe coral bleaching events during the last two decades. The impacts of coastal development, destructive fishing and the expansion of tourism on coral reefs are documented [12,40]. The coral communities at Ko Losin (West), Ko Losin (East) and Ko Ngam Noi are interesting due to their high percentages of live coral cover and the fact that the dominant corals of these reef sites are several species of *Acropora*, which are susceptible to abnormal high-temperature-driven coral bleaching [7,45]. The coral communities at the study sites of Ko Losin are in relatively deep water, which may have protected them from high temperatures during the severe coral bleaching events in 1998 and 2010. Some *Acropora* corals also showed a high degree of bleaching but they did not die after bleaching. Intensive studies on ocean currents and other related issues of physical oceanography are required for understanding high resistance to bleaching events. Protection of the coral communities at Ko Losin from negative impacts of human activities, particularly fishing, boat anchoring and diving, is urgently needed to enhance coral reef resilience in the Gulf of Thailand.

The density of juvenile corals in the Gulf of Thailand is usually lower compared to that of other reef sites in the Indo-Pacific region [46]. Therefore, the coral communities in the Gulf of Thailand can maintain their community structures through the survival of resistant and/or tolerant coral species. The results of this study suggest that highly resistant and tolerant coral species at Ko Losin, Ko Ngam Noi and Ko Kula play a major role in the high resilience potential of coral communities after coral bleaching events. The *Acropora* communities at Ko Ngam Noi, Mu Ko Chumphon National Park, are particularly important to the high resilience potential of nearshore reef sites. These coral communities may provide larval supply to nearshore reefs along the Western Gulf of Thailand through the connecting sea surface current in the Gulf of Thailand [47].

The poor coral condition at Ko Ngam Yai and the high percentage of dead corals at Ko Kula in Mu Ko Chumphon National Park imply the need for urgent investigation on how to restore these reef sites. The densities of juvenile corals at Ko Ngam Yai and Ko Kula from this study were relatively high compared to those of other reef sites in the Gulf of Thailand. The dominant juvenile corals at Ko Ngam Yai were *Pocillopora*, *Porites*, *Favites* and *Pavona*, whereas the dominant juvenile corals at Ko Kula were *Fungia*, *Porites* and *Pachyseris*. Enhancing the survival rates of juvenile corals is crucial for coral recovery following bleaching events [38]. Sediment loaded from coastal development and tourism impacts should be carefully mitigated for passive coral reef restoration. A high diversity of healthy corals in a coral reef ecosystem is an important factor for enhancing reef resilience potential because it occupies the reef substrates and inhibits the settlement of other benthic organisms that are coral competitors [17]. The coral communities at Ko Kula and Ko Ngam Yai also require an adequate supply of coral larvae from other coral reefs in the Gulf of the Thailand to enhance their coral diversity.

The density of juvenile corals recorded in our study was 0.89–3.73 colonies/m^2, which is comparable to that of the Palk Bay reef in the northern Indian Ocean [17] but is much lower than that of several reef sites in the Indo-Pacific region, in which the juvenile coral density at some reef sites was over 50 colonies/m^2 [48,49]. Variation in the juvenile coral density between the study sites of Mu

Ko Chumphon and Ko Losin was obviously shown in this study. Several factors may influence this spatial variation in juvenile coral density, such as larval supply from the parent reef, larval mortality, reef connectivity, settlement and post-settlement mortality, grazing and sedimentation [50,51]. The density of the juvenile corals at Ko Losin (West), Ko Losin (East), Ko Ngam Noi and Ko Kula was not dependent on the live coral cover of adult coral colonies in a reef. Moreover, the *Acropora* communities at Ko Losin and Ko Ngam Noi had no juvenile corals in their communities.

This study shows that several coral reefs at Ko Losin and Mu Ko Chumphon in the south of Thailand had high resilience potential to coral bleaching events and anthropogenic disturbances because of their survival rates, although they had relatively low densities of juvenile corals. We suggest that Ko Losin should be established as a marine protected area under Thai laws to protect the healthy corals as well as to provide coral larvae to other coral reefs in the Gulf of Thailand. The results from this study also imply that Mu Ko Chumphon National Park should implement its management plans properly to enhance coral recovery at Ko Ngam Yai and Ko Kula. Resilience-based management may be applied to support natural processes that promote the resistance and recovery of corals [43]. The promotion of marine ecotourism can protect coral communities at tourist destinations as well as keep the tourist numbers below the carrying capacity of the reef sites. Other measures to enhance the resistance of corals during bleaching events and appropriate coral restoration projects should be also considered. The field shading experiments that were carried out on coral communities of Ko Ngam Noi should be applied at other reef sites to protect corals during bleaching periods [41].

Author Contributions: All of the authors collected data; M.S. and T.Y. conceived the idea; M.S., T.Y., C.C., S.P. and W.K. analysed the data and wrote the manuscript.

Funding: This research was funded by Thailand Science Research and Innovation (TSRI), National Science and Technology Development Agency (NSTDA) and a budget for research promotion from the Thai Government to Ramkhamhaeng University.

Acknowledgments: We thank Loke Ming Chou and Danwei Huang, the subject editors of this special issue, for encouraging us to analyze the coral reef data from Thailand. We are most grateful to the staff of Marine National Park Operation Center Chumphon, Department of National Park, Wildlife and Plant Conservation, Save Our Sea Association (SOS) and Marine Biodiversity Research Group, Faculty of Science, Ramkhamhaeng University, for their support and assistance in the field. This research was funded by the Thailand Science Research and Innovation (TSRI), National Science and Technology Development Agency (NSTDA) and a budget for research promotion from the Thai Government to Ramkhamhaeng University. We also thank three anonymous reviewers for providing valuable comments and editing the manuscript.

Conflicts of Interest: The authors declare no conflict of interest.

References

1. Burke, L.; Reytar, K.; Spalding, M.; Perry, A. *Reefs at Risk Revisited*; World Resources Institute: Washington, DC, USA, 2011; p. 115.
2. Mora, C.; Graham, N.A.J.; Nyström, M. Ecological limitations to the resilience of coral reefs. *Coral Reefs* **2016**, *35*, 1271–1280. [CrossRef]
3. Harvell, C.D.; Jordán-Dahlgren, E.; Merkel, S.; Raymundo, L.; Smith, G.; Weil, E.; Willis, B. Coral disease, environmental drivers, and the balance between coral and microbial associates. *Oceanography* **2007**, *20*, 172–195. [CrossRef]
4. Guillemot, N.; Chabanet, P.; Le Pape, O. Cyclone effects on coral reef habitats in New Caledonia (South Pacific). *Coral Reefs* **2010**, *29*, 445–453. [CrossRef]
5. Hughes, T.P.; Graham, N.A.; Jackson, J.B.; Mumby, P.J.; Steneck, R.S. Rising to the challenge of sustaining coral reef resilience. *Trends. Ecol. Evol.* **2010**, *25*, 633–642. [CrossRef] [PubMed]
6. Pandolfi, J.M.; Connolly, S.R.; Marshall, D.J.; Cohen, A.L. Projecting coral reef futures under global warming and ocean acidification. *Science* **2011**, *333*, 418–422. [CrossRef] [PubMed]
7. Yeemin, T.; Pengsakun, S.; Yucharoen, M.; Klinthong, W.; Sangmanee, K.; Sutthacheep, M. Long-term decline in Acropora species at Kut Island, Thailand, in relation to coral bleaching events. *Mar. Biodiv.* **2013**, *43*, 23–29. [CrossRef]

8. Palumbi, S.R.; Barshis, D.J.; Traylor-Knowles, N.; Bay, R.A. Mechanisms of reef coral resistance to future climate change. *Science* **2014**, *344*, 895–898. [CrossRef] [PubMed]

9. Samsuvan, W.; Yeemin, T.; Sutthacheep, M.; Pengsakun, S.; Putthayakool, J.; Thummasan, M. Diseases and compromised health states of massive *Porites* spp. in the Gulf of Thailand and the Andaman Sea. *Acta Oceanol. Sin.* **2019**, *38*, 118–127. [CrossRef]

10. Valentine, J.F.; Heck, K.L., Jr. Perspective review of the impacts of overfishing on coral reef food web linkages. *Coral Reefs* **2005**, *24*, 209–213. [CrossRef]

11. Guest, J.R.; Low, J.; Tun, K.; Wilson, B.; Ng, C.; Raingeard, D.; Ulstrup, K.E.; Tanzil, J.T.I.; Todd, P.A.; Toh, T.C.; et al. Coral community response to bleaching on a highly disturbed reef. *Sci. Rep.* **2016**, *6*, 20717. [CrossRef] [PubMed]

12. Yeemin, T.; Pengsakun, S.; Yucharoen, M.; Klinthong, W.; Sangmanee, K.; Sutthacheep, M. Long-term changes of coral communities under stress from sediment. *Deep-Sea Res. II* **2013**, *96*, 32–40. [CrossRef]

13. Heery, E.C.; Hoeksema, B.W.; Browne, N.K.; Reimer, J.D.; Ang, P.O.; Huang, D.; Friess, D.A.; Chou, L.M.; Loke, L.H.L.; Saksena-Taylor, P.; et al. Urban coral reefs: Degradation and resilience of hard coral assemblages in coastal cities of East and Southeast Asia. *Mar. Pollut. Bull.* **2018**, *135*, 654–681. [CrossRef] [PubMed]

14. Hughes, T.P.; Baird, A.H.; Bellwood, D.R.; Card, M.; Connolly, S.R.; Folke, C.; Grosberg, R.; Hoegh-Guldberg, O.; Jackson, J.B.C.; Kleypas, J. Climate change, human impacts, and the resilience of coral reefs. *Science* **2003**, *301*, 929–933. [CrossRef] [PubMed]

15. Baker, A.C.; Glynn, P.W.; Riegl, B. Climate change and coral reef bleaching: An ecological assessment of long-term impacts, recovery trends and future outlook. *Estuar. Coast. Shelf Sci.* **2008**, *80*, 435–471. [CrossRef]

16. Norström, A.V.; Nystrom, M.; Lokrantz, J.; Folke, C. Alternative states on coral reefs: Beyond coral-macroalgal phase shifts. *Mar. Ecol. Prog. Ser.* **2009**, *376*, 295–306. [CrossRef]

17. Manikandan, B.; Ravindran, J.; Vidya, P.J.; Shrinivasu, S.; Manimurali, R.; Paramasivam, K. Resilience potential of an Indian Ocean reef: An assessment through coral recruitment pattern and survivability of juvenile corals to recurrent stress events. *Environ. Sci. Pollut. Res.* **2017**, *24*, 13614–13625. [CrossRef] [PubMed]

18. Anthony, K.R.; Marshall, P.A.; Abdulla, A.; Beeden, R.; Bergh, C.; Black, R.; Eakin, C.M.; Game, E.T.; Gooch, M.; Graham, N.A. Operationalizing resilience for adaptive coral reef management under global environmental change. *Glob. Chang. Biol.* **2015**, *21*, 48–61. [CrossRef] [PubMed]

19. Mumby, P.J.; Hastings, A.; Edwards, J.G. Thresholds and the resilience of Caribbean coral reefs. *Nature* **2007**, *450*, 98–101. [CrossRef] [PubMed]

20. McClanahan, T.R.; Donner, S.D.; Maynard, J.A.; MacNeil, M.A.; Graham, N.A.J.; Maina, J.; Baker, A.C.; Alemu, I.J.B.; Beger, M.; Campbell, S.J. Prioritizing key resilience indicators to support coral reef management in a changing climate. *PLoS ONE* **2012**, *7*, e42884. [CrossRef] [PubMed]

21. Graham, N.A.; Bellwood, D.R.; Cinner, J.E.; Hughes, T.P.; Norström, A.; Nyström, M. Managing resilience to reverse phase shifts in coral reefs. *Front. Ecol. Environ.* **2013**, *11*, 541–548. [CrossRef]

22. Salm, R.V.; Smith, S.E.; Llewellyn, G. Mitigating the impact of coral bleaching through marine protected area design. In *Coral Bleaching: Causes, Consequences and Response*; Schuttenberg, H.Z., Ed.; University of Rhode Island: Kingston, RI, USA, 2001; pp. 81–88.

23. West, J.M.; Salm, R.V. Resistance and resilience to coral bleaching: Implications for coral reef conservation and management. *Conserv. Biol.* **2003**, *17*, 956–967. [CrossRef]

24. Obura, D.; Grimsditch, G. *Resilience Assessment of Coral Reefs: Assessment Protocol for Coral Reefs, Focusing on Coral Bleaching and Thermal Stress*; IUCN: Gland, Switzerland, 2009; p. 70.

25. Maynard, J.A.; Marshall, P.A.; Johnson, J.E.; Harman, S. Building resilience into practical conservation: Identifying local management responses to global climate change in the southern Great Barrier Reef. *Coral Reefs* **2010**, *29*, 381–391. [CrossRef]

26. Beeden, R.; Maynard, J.; Johnsone, J.; Dryden, J.; Kininmonth, S. No-anchoring areas reduce coral damage in an effort to build resilience in Keppel Bay, southern Great Barrier Reef. *Aust. J. Environ. Manag.* **2014**, *21*, 311–319. [CrossRef]

27. Bachtiar, I.; Suharsono; Damar, A.; Zamani, N.P. Practical Resilience Index for Coral Reef Assessment. *Ocean Sci. J.* **2019**, *54*, 117–127. [CrossRef]

28. Hughes, T.P.; Baird, A.H.; Dinsdale, E.A.; Harriott, V.J.; Moltschaniwskyj, N.A.; Pratchett, M.S.; Tanner, J.E.; Willis, B.L. Detecting regional variation using meta-analysis and large-scale sampling: Latitudinal patterns in recruitment. *Ecology* **2002**, *83*, 436–451. [CrossRef]

29. Doropoulos, C.; Ward, S.; Roff, G.; González-Rivero, M.; Mumby, P.J. Linking demographic processes of juvenile corals to benthic recovery trajectories in two common reef habitats. *PLoS ONE* **2015**, *10*, e0128535. [CrossRef] [PubMed]

30. Yucharoen, M.; Yeemin, T.; Casareto, B.E.; Suzuki, Y.; Samsuvan, W.; Sangmanee, K.; Klinthong, W.; Pengsakun, S.; Sutthacheep, M. Abundance, composition and growth rate of coral recruits on dead corals following the 2010 bleaching event at Mu Ko Surin, the Andaman Sea. *Ocean Sci. J.* **2015**, *50*, 307–315. [CrossRef]

31. Bramanti, L.; Edmund, P.J. Density-associated recruitment mediates coral population dynamics on a coral reef. *Coral Reefs* **2016**, *35*, 543–553. [CrossRef]

32. Yeemin, T.; Saenghaisuk, C.; Yucharoen, M.; Klinthong, W.; Sutthacheep, M. Impact of the 2010 coral bleaching event on survival of juvenile coral colonies in the Similan Islands, on the Andaman Sea coast of Thailand. *Phuket Mar. Biol. Cent. Res. Bull.* **2012**, *71*, 93–102.

33. Rotha, F.; Saalmann, F.; Thomson, T.; Coker, D.J.; Villalobos, R.; Jones, B.H.; Wild, C.; Carvalho, S. Coral reef degradation affects the potential for reef recovery after disturbance. *Mar. Environ. Res.* **2018**, *142*, 48–58. [CrossRef] [PubMed]

34. Graham, N.A.J.; Nash, K.L.; Kool, J.T. Coral reef recovery dynamics in a changing world. *Coral Reefs* **2011**, *30*, 283–294. [CrossRef]

35. Richmond, R.H. Reproduction and recruitment in corals: Critical links in the persistence of reefs. In *Life and Death of Coral Reefs*; Birkeland, C., Ed.; Chapman & Hall: New York, NY, USA, 1997; pp. 175–197.

36. Kuffner, I.B.; Walters, L.J.; Becerro, M.A.; Paul, V.J.; Ritson-Williams, R.; Beach, K.S. Inhibition of coral recruitment by macroalgae and cyanobacteria. *Mar. Ecol. Prog. Ser.* **2006**, *323*, 107–117. [CrossRef]

37. Perez, K.; Rodgers, K.S.; Jokiel, P.L.; Lager, C.V.; Lager, D.J. Effects of terrigenous sediments on settlement and survival of the reef coral. *Pocillopora damicornis*. *PeerJ* **2014**, *2*, e387. [CrossRef] [PubMed]

38. Shlesinger, T.; Loya, Y. Recruitment, mortality, and resilience potential of scleractinian corals at Eilat, Red Sea. *Coral Reefs* **2016**, *35*, 1357–1368. [CrossRef]

39. Yeemin, T.; Mantachitra, V.; Plathong, S.; Nuclear, P.; Klinthong, W.; Sutthacheep, M. Impacts of coral bleaching, recovery and management in Thailand. In Proceedings of the 12th International Coral Reef Symposium, Cairns, Australia, 9–13 July 2012; p. 5.

40. Sutthacheep, M.; Pengsakun, S.; Yucharoen, M.; Klinthong, W.; Sangmanee, K.; Yeemin, T. Impacts of the mass coral bleaching events in 1998 and 2010 on the western Gulf of Thailand. *Deep-Sea Res. II* **2013**, *96*, 25–31. [CrossRef]

41. Yeemin, T. Summary of coral bleaching from 2015 to 2017 in Thailand. In *Status of Coral Reefs in East Asian Seas Region*; Kimura, T., Tun, K., Chou, L.M., Eds.; Ministry of the Environment of Japan and Japan Wildlife Research Center: Tokyo, Japan, 2018; pp. 25–28.

42. Holling, C.S. Resilience and stability of ecological systems. *Annu. Rev. Ecol. Systemat.* **1973**, *4*, 1–23. [CrossRef]

43. Mcleod, E.; Anthony, K.R.N.; Mumby, P.J.; Maynard, J.; Beeden, R.; Graham, N.A.J.; Heron, S.F.; Hoegh-Guldberg, O.; Jupiter, S.; MacGowan, P.; et al. The future of resilience-based management in coral reef ecosystems. *J. Environ. Manag.* **2019**, *233*, 291–301. [CrossRef] [PubMed]

44. Veron, J.E.N. *Corals of the World*; Australian Institute of Marine Science: Townsville, Australia, 2000; Volume 1–3.

45. Loya, Y.; Sakai, K.; Yamazato, K.; Nakano, Y.; Sambali, H.; van Woesik, R. Coral bleaching: The winners and the losers. *Ecol. Lett.* **2001**, *4*, 122–131. [CrossRef]

46. Yeemin, T.; Saenghaisuk, C.; Sutthacheep, M.; Pengsakun, S.; Klinthong, W.; Saengmanee, K. Conditions of coral communities in the Gulf of Thailand: A decade after the 1998 bleaching event. *Galaxea J. Coral Reef Stud.* **2009**, *11*, 207–217. [CrossRef]

47. Sojisuporn, P.; Morimoto, A.; Yanagi, T. Seasonal variation of sea surface current in the Gulf of Thailand. *Coast. Mar. Sci.* **2010**, *34*, 1–12.

48. Sheppard, C.C.R.; Harris, A.; Sheppard, A.L.S. Archipelago-wide coral recovery patterns since 1998 in the Chagos Archipelago, Central Indian Ocean. *Mar. Ecol. Prog. Ser.* **2008**, *362*, 109–117. [CrossRef]

49. Roth, M.S.; Knowlton, N. Distribution, abundance and microhabitat characterization of small juvenile corals at Palymra Atoll. *Mar. Ecol. Prog. Ser.* **2009**, *376*, 133–142. [CrossRef]
50. Edmunds, P.J.; McIlroy, S.E.; Adjeroud, M.; Ang, P.; Bergman, J.L.; Carpenter, R.C.; Coffroth, M.A.; Fujimura, A.; Hench, J.; Holbrook, S.J.; et al. Critical information gaps impeding understanding of the role of larval connectivity among coral reef islands in an era of global change. *Front. Mar. Sci.* **2018**, *5*, 290. [CrossRef]
51. Sutthacheep, M.; Sakai, K.; Yeemin, T.; Pensakun, S.; Klinthong, W.; Samsuvan, W. Assessing Coral Reef Resilience to Climate Change in Thailand. *RIST* **2018**, *1*, 22–34.

Review

Coral Reef Resilience in Taiwan: Lessons from Long-Term Ecological Research on the Coral Reefs of Kenting National Park (Taiwan)

Shashank Keshavmurthy [1,†], Chao-Yang Kuo [1,†], Ya-Yi Huang [1], Rodrigo Carballo-Bolaños [1,2,3], Pei-Jei Meng [4,5,*], Jih-Terng Wang [6,*] and Chaolun Allen Chen [1,2,3,7,8,*]

[1] Biodiversity Research Center, Academia Sinica, Taipei 11529, Taiwan; coralresearchtaiwan@gmail.com (S.K.); cykuo.tw@gmail.com (C.-Y.K.); lecy.yhuang@gmail.com (Y.-Y.H.); rodragoncar@gmail.com (R.C.-B.)
[2] Biodiversity Program, Taiwan International Graduate Program, Academia Sinica, Taipei 11529, Taiwan
[3] Department of Life Science, National Taiwan Normal University, Taipei 10610, Taiwan
[4] National Museum of Marine Biology/Aquarium, Pintung 94450, Taiwan;
[5] Insititute of Marine Biodiversity and Evolution, National Dong-Hua University, Hualien 97447, Taiwan
[6] Department of Biotechnology, Tajen University, Pingtung 90741, Taiwan
[7] Institute of Oceanography, National Taiwan University, Taipei 10617, Taiwan
[8] Department of Life Science, Tunghai University, Taichung 40704, Taiwan
[*] Correspondence: pjmeng@nmmba.gov.tw (P.-J.M.); jtwtaiwan@gmail.com (J.-T.W.); cac@gate.sinica.edu.tw (C.A.C.)
[†] These two authors contributed equally to the work described.

Received: 20 September 2019; Accepted: 28 October 2019; Published: 31 October 2019

Abstract: Coral reefs in the Anthropocene are being subjected to unprecedented levels of stressors, including local disturbances—such as overfishing, habitat destruction, and pollution—and large-scale destruction related to the global impacts of climate change—such as typhoons and coral bleaching. Thus, the future of corals and coral reefs in any given community and coral-Symbiodiniaceae associations over time will depend on their level of resilience, from individual corals to entire ecosystems. Herein we review the environmental settings and long-term ecological research on coral reefs, based on both coral resilience and space, in Kenting National Park (KNP), Hengchun Peninsula, southern Taiwan, wherein fringing reefs have developed along the coast of both capes and a semi-closed bay, known as Nanwan, within the peninsula. These reefs are influenced by a branch of Kuroshio Current, the monsoon-induced South China Sea Surface Current, and a tide-induced upwelling that not only shapes coral communities, but also reduces the seawater temperature and creates fluctuating thermal environments which over time have favoured thermal-resistant corals, particularly those corals close to the thermal effluent of a nuclear power plant in the west Nanwan. Although living coral cover (LCC) has fluctuated through time in concordance with major typhoons and coral bleaching between 1986 and 2019, spatial heterogeneity in LCC recovery has been detected, suggesting that coral reef resilience is variable among subregions in KNP. In addition, corals exposed to progressively warmer and fluctuating thermal environments show not only a dominance of associated, thermally-tolerant *Durusdinium* spp. but also the ability to shuffle their symbiont communities in response to seasonal variations in seawater temperature without bleaching. We demonstrate that coral reefs in a small geographical range with unique environmental settings and ecological characteristics, such as the KNP reef, may be resilient to bleaching and deserve novel conservation efforts. Thus, this review calls for conservation efforts that use resilience-based management programs to reduce local stresses and meet the challenge of climate change.

Keywords: Taiwan; coral reef; marine national park; nuclear power plant; community dynamics; Symbiodiniaceae; long-term ecological data

---

1. Introduction

1.1. Coral Reef Ecosystems and the Impacts of Environmental Change

Coral reefs are one of the world's most diverse and productive marine ecosystems, providing essential goods and services for millions of people. A remarkably high diversity of species and interactions makes this ecosystem particularly sophisticated and sensitive to human-induced perturbations [1]. Overfishing, pollution and habitat destruction are among the most common local stressors altering coral reef ecosystem dynamics (reviewed in [1–4]). In addition, the increases in seawater temperature and ocean acidification are considered to be the two major global stressors responsible for the worldwide degradation of coral reef health (reviewed in [3]). Seawater temperatures 0.5–1.0 °C above summer average persisting for days to weeks could result in disruption of mutualistic relationship between coral hosts and their symbiotic algae (also known as zooxanthellae, family Symbiodiniaceae), resulting in "coral bleaching"; if this persists for months, it can result in mass coral mortality [5]. Several major severe bleaching episodes have been recorded since 1979. Among them, a 1998 event had the most devastating known effect, impacting over 75% of reefs worldwide, and wiping out nearly 16% of them [6]. In addition, between 2014 and 2017, "back-to-back" thermal anomalies occurred on the northeastern coast of Australia, causing massive coral bleaching in the north and mid sections of the Great Barrier Reef; this is arguably the worst-ever bleaching in the history of the GBR [7,8] and other parts of Australia [9,10]. Similar global-scale coral bleaching events (GCBE) that result in high coral mortality, the rapid decline of reef structures, and unprecedented environmental impacts have also been reported in the Indian [11,12], Pacific [13–15], and Atlantic Oceans [16,17]. Scientists have therefore concluded that the 2014–2017 GCBE represents the first multi-year, global-scale coral bleaching event to cause bleaching and mortality two or more times over the 3-year event [18].

Ocean acidification—the ongoing decrease in the pH of the ocean caused by the uptake of anthropogenic carbon dioxide (CO_2)—is, on the other hand, decreasing the calcification, reducing coral growth, and limiting reef development. There appears to be no chance of maintaining atmospheric CO_2 concentrations under 450 ppm [19] and limiting global temperature increase to less than 2 °C [20] by the end of the century; however, corals and coral reef communities at low and high latitudes might demonstrate exceptional acclimatization and adaptation capacities to survive the future environmental changes [21], although it has been suggested that most species will fail to develop mechanisms to survive future conditions, and a worldwide decline in coral reefs now seems inevitable [3]. Recent reports have shown that recurrent and prolonged thermal anomalies above the threshold limit corals' resistance to stress [22] and leads to increased bleaching, mass destruction of coral reefs, and the loss of many coral species [7,8]. The latest IPCC reports show an imminent threat to tropical coral reefs as soon as 2030 if carbon emissions continue to increase and subsequently increase average seawater temperatures. It was concluded at 1.5 °C global heating, the world will lose 70%–90% of its coral reefs, but at 2 °C, virtually all of the world's coral reefs will be lost [19].

In addition to driving more intensive mass coral bleaching, warming oceans will also enhance the destructive potential of tropical storms, including typhoons in the West Pacific, hurricanes in the Caribbean, and cyclones in the South Pacific and Indian Oceans, by either increasing their frequencies [23] or intensities [24], although other studies suggest that the global frequency of cyclones might remain stable or even decrease by up to 40% under greenhouse conditions by the end of this century [25–27]. Despite these projections, future impacts remain debatable. The future climate might, in turn, be influenced by some of the most severe ecological impacts on coral reefs through direct physical disturbances, turbidity, sedimentation, or salinity changes that would result from the destruction of reef structures [28]. As a consequence, the diversity and biomass of fish and other fauna that require corals for shelter or food will be dramatically reduced [29].

1.2. Coral Reef Resilience under the Impacts of Environmental Change

Resilience, a theory introduced to describe how ecosystems respond to disturbances [20], has recently been applied to coral reef ecosystems to examine how organisms respond to and interact as the result of local and global stresses. Many studies [30–35] including [36] have developed different definitions of resilience when applied to coral reef ecosystems including that the resilience is influenced by many stochastic factors [36]. In other words, stochastic resilience or ecosystem resilience is the capacity of an ecosystem to overcome disturbances and reorganize to maintain original fundamental state [36–38]. In case of coral reefs, the disturbances vary in terms of time and space with different intensities. As a result, depending on the location and local environmental factors, resilience of a given reef and its coral communities will vary. Resilience-based management has been proposed as a realistic model to predict which sustainability measures can be achieved in coral reefs in the face of ocean warming and acidification and various local disturbances, and this strategy will help us set achievable goals for regional and local-scale management programs [39]. Moreover, beyond the ecosystem level, resilience also covers the overall ability of individuals, populations, or communities to respond positively after disturbance and restore some part of their original state [4]. For example, individual corals can show physiological resilience via survival, sustained growth, reproduction, and/or by shuffling their symbionts towards more thermally-tolerant genera and species (see the review by Carballo-Bolaños et al. in this issue). Coral populations can re-populate by recruiting new individuals, and communities can re-organise ecosystem traits such as productivity, diversity, trophic linkages, and sustained biomass through shifts in species composition [reviewed in 4]. In this review, we adopted the broad definition of resilience [4] to examine the resilience of coral-Symbiodiniaceae associations and coral communities to the long-term disturbances in KNP.

1.3. Coral-Symbiodiniaceae Associations Play a Key Role in Coral Resilience to Thermal Stress

Coral-Symbiodiniaceae associations are the crux of coral health and functioning, even though corals are multi-bionts [40]. One of the most important factors of this association is its resilience to seawater temperature anomalies. Coral-Symbiodiniaceae resistance mechanisms to temperature stress depend on the combination of mechanisms involving the coral host and/or its Symbiodiniaceae partners and whether the relationship that the coral host has with Symbiodiniaceae is specific or flexible [41–43]. Corals are known to associate with a wide range of Symbiodiniaceae genera. There are nine genera in Symbiodiniaceae, and each genus has its own characteristics that help corals survive in a wide range of environmental niches [44]. Studies have shown that symbiosis between coral hosts and different Symbiodiniaceae genera contribute to the divergence in their thermal tolerance under different environmental conditions [42,45].

For example, stress-tolerant *Durusdinium trenchii* has proven to be heat-tolerant, whereas *Cladocopium* species are the most sensitive to stress [41,46–48]. *Durusdinium*-associated corals are also known to inhabit reef environments that experience large surface seawater temperature fluctuations [42,49,50] and possess better survival chances under heat-treatment experiments [51]. Symbiodiniaceae diversity thus provides a mechanism for corals to adapt and/or acclimatize to changing environmental conditions. Since environmental factors vary both spatially and temporally, even at a micro-geographic scale, differences in Symbiodiniaceae diversity within a single host or among multiple species are important and contribute to coral resilience.

1.4. Coral Reefs in Taiwan, with a Focus on Kenting Coral Reefs

Taiwan, a continental island with several offshore islets, is located at the centre or junction of the Philippine-Japan island arc; the Tropic of Cancer, runs through the middle of Taiwan (Figure 1). Scleractinian coral occurrence and distribution in Taiwan is influenced by sea surface currents and seawater temperatures; it is found in patchy non-reefal coral communities, similar to a high-latitudinal environment, and occupies the Penghu Archipelago and northern, northeastern, and rocky eastern coasts of Taiwan [52–54]. The marine environmental conditions in southern Taiwan—facing the junction of the Pacific Ocean, Bashi Channel, and South China Sea—are influenced by a branch of the Kuroshio Current (KC) and the South China Sea Surface Current (SCSSC) (Figure 1). Thus, tropical fringing reefs developed along the coast of the Hengchun Peninsula, with about 300 coral species described [52–54].

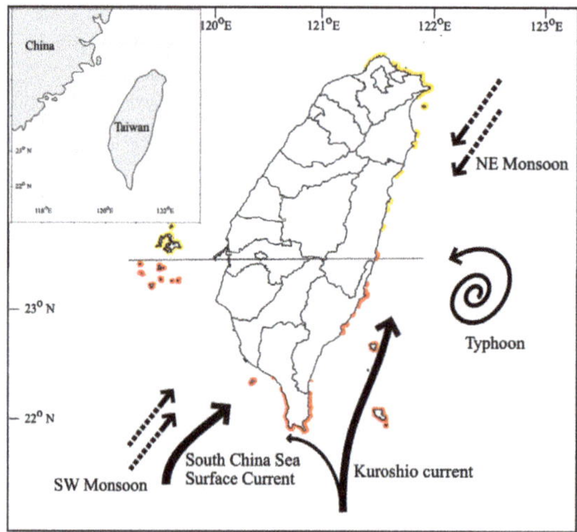

Figure 1. The distribution of corals in Taiwan, including non-reefal communities (yellow) and tropical coral reefs (red) divided roughly by the Tropic of Cancer, and the main climate factors, including the Kuroshio Current, typhoons, and northeast and southwest monsoons. The two types of coral communities are divided by the Tropic of Cancer. The information in the map was adapted from (Dai, C.-F. 2018). Taiwan map was downloaded from free to use Taiwan Map Store (https://whgis.nlsc.gov.tw/English/0-1Introduction.aspx).

In 1982, Kenting National Park (KNP), the first national park in Taiwan, was established to manage and conserve the uplifted reef landscape as well as the modern coral reef system (Figure 2). In addition, a nuclear power plant (NPP) that started operating in 1985 discharges heated seawater into Nanwan, KNP (Figure 2). Biological surveys and environmental monitoring have been conducted in KNP since the late 1970s to collect baseline data to manage KNP and monitor the environmental impact of NPP during construction and follow-up operations, particularly the thermal stress on the reef adjacent to the heated-water outlet (OL); data collection was interrupted from time to time due to shortages in funding or termination of monitoring projects (Table 1). Nevertheless, this long-term research effort (> 30 years) collected enough data for us to assess spatial and temporal changes in Taiwan's coral reef ecosystems (Figure 2). However, similar to reefs from most of the tropical waters [55–58], the reefs in KNP have been subjected to human disturbances—including overfishing, habitat destruction, and sewage discharge—due to a growing population and poor management [59–62]. Combined with the synergistic effects from climate change, there has been a declining trend in KNP's living coral cover [60,63].

Figure 2. Depth contours in Nanwan, Kenting National Park, and coral communities at six sites and two time points, showing the status of coral communities through time. The sites include Wanlitung (WLT) on the west coast of Hengchun Peninsula, NPP Outlet (OL) and Houbihu (HBH) on the west coast of Nanwan, Tiaoshi (TS) and Siangjiaowan (SJW) on the east coast of Nanwan, and Longkeng (LK) on the east coast of Hengchun Peninsula. The solid and dotted arrow line indicates the direction of tidal flow patterns in flows and eddies. Depth contour figure was adapted from Lee, H.-J. etc. 1999a.

In this paper, we review the long-term studies of (1) environmental settings of KNP coral reefs, focusing on the cooling effect of tide-induced upwelling; (2) spatial and temporal dynamics of benthic communities and how they respond to large-scale disturbances, such as coral bleaching and typhoons; and (3) Symbiodiniaceae diversity and the responses of KNP corals to thermal-induced bleaching. By synthesizing the long-term ecological research data on coral reefs in KNP, we show that, by introducing resilience-based management programs to reduce local stresses, coral distribution along a small geographic range with unique environmental settings and Symbiodiniaceae diversity can be potentially resilient to climate change.

This section can be summarized as; (1) Coral reefs are facing increasing pressure from natural and anthropogenic disturbances and (2) Resilience of corals to such disturbances will depend on local-scale management programs, physiological resilience of individual species, recruitment potential, etc.

Table 1. Summary of the historical dataset conducting National Park-wide qualitative survey since 1986.

Year	Sites	Survey Method	Survey Area per Transect	Number of Replicates at Each Site	Survey Depth	Methods	Identification Level	Reference
1986–1987, 1998–1999	Wanlitung (WLT) Hungchai (HC) Leidashih (LDS) Houbihu (HBH) Tiaoshi (TS) Siangjiaowan (SJW)	Line intercept transect	10 m	25	3–23	A transect tape was placed perpendicular to the coast and extended seaward from 3 m depth to the reef edge at 25 m depth. A 10 m metal chain was placed parallel to the transect at 15 m intervals.	Species for corals Total algae	Dai, C.-F. (1988); Dai, C.-F. etc. (1998, 1999)
2003–2005 2008–2014 2016, 2018	Howan (HW) Wanlitung (WLT) Hungchai (HC) Leidashih (LDS) Houbihu (HBH) NPP Inlet (IL) Tiaoshi (TS) Siangjiaowan (SJW) Longkeng (LK) Jialeshuei (JLS)	30 m × 0.25 m belt transects except 20 m × 0.25 m in TS	7.5 m^2 and 5 m^2 in TS	3, except 9 in TS	5–10	Three or nine (at Tiaoshi) permanent belt transects were established along depth contours between 5 and 10 m depth at each site. Benthic organisms were quantified using 25 × 25 cm^2 photo-quadrats. The percent cover of the benthic categories was determined using Coral Count with Excel Extensions software (Kohler, K.E. 2016), with 30 random points per quadrat. Surveys were conducted between March and May each year.	Species for corals (2003–2005, 2011, 2014); Genus for corals (2008–2010, 2012–2013); Morphology for scleractinian corals (2016); Genus and morphology for scleractinian corals (2018) Macroalgae Turf algae	Kuo, C.-Y. etc. (2012); Shao, K.-T. etc. (2002); Fang, L.-S. etc. (2003–2006); Wang, W.-H. etc. (2007); Ho, M.-J. etc. (2016); Ho, P.-H. etc. (2008–2011); Chen, C.A. etc. (2012–2014, 2016, 2018); Kuo, C.-Y. etc. (2007)

2. Environmental Settings of Coral Reef in KNP with Focus on Tide-Induced Upwelling

KNP is located on Hengchun Peninsula in southernmost Taiwan and receives seasonal influence from a branch of KC and SCSSC; the park can be divided into three geographic sub-regions, namely the west coast of Hengchun Peninsula facing the Taiwan Strait; the east coast of Hengchun Peninsula adjacent to the Pacific Ocean; and Nanwan, a semi-enclosed bay between two capes facing the Bashi Channel that connects the Pacific Ocean and Taiwan Strait (Figure 1). Distance between the east cape, known as Eluanbi, protruding farther south and the west cape, Maobitou, is about 14 km. The semi-enclosed basin of Nanwan is characterized by a zonally elongated seamount partially blocking the bay mouth. The seamount reaches up to 50 m below the sea surface (Figure 2). Toward the west side of Nanwan, there is almost no continental shelf. On the contrary, the shallow continental shelf shoreward of 80 m isobath is about 4 km wide, with isobaths running more or less parallel to the coastline at the east side of Nanwan. With this unique geomorphologic setting, the deeper portion of Nanwan forms an arc-shaped channel open at both ends between the seamount to the south and landmasses to the north [61–63].

By applying acoustic Doppler current profiler surveys, moored measurements, and numerical modelling, it has been shown that the headlands on either side of the Nanwan generate strong tidally-induced upwelling (TIU) within the bay during each phase of the tide [64,65]. Considerable difference in size between the flood and ebb eddies were observed following the geometry of the region. The entire Nanwan basin is filled by the flood eddy, while the western and central regions are filled by the ebb eddy. Eventually, the upwelling occurs within each eddy, causing two temperature drops per tidal cycle in western and central Nanwan, and only one drop in the eastern part (Figure 3) [64,65].

The significant temperature drops caused by TIU (Figure 3) might not only play an important role in shaping the coral community structure in different sub-regions of KNP, but also have a crucial cooling effect on corals in the Nanwan that resists the rising seawater temperature, both locally and globally. Analysis of in situ hourly temperature data showed a spatial heterogeneity in response to TIU at different sub-regions in KNP (Figure 3). No TIU was detected on the west coast of Hengchun Peninsula; thus, seawater temperature is constantly stable, with small fluctuations through the year (Figure 3a,b). In contrast, within the Nanwan and the east coast of Hengchun Peninsula, seawater temperature responds in concordance, but with different amplitudes, to the TIU, following the moon phases (Figure 3c–j). In west Nanwan, where the OL is located, large amplitudes were observed around new moons, with a maximum temperature fluctuation of 4.08 °C and 4.89 °C at 2 m and 7 m deep in the summer, respectively (Figure 3c–f), providing a significant cooling effect to remove thermal stress on coral reefs near the OL (Table 2). Towards east Nanwan and the corner of east Hengchun Peninsula, the effect of temperature drop by TIU is reduced in the summer, as the model predicted [64,65], but remains relatively strong in the winter (Figure 3g–j).

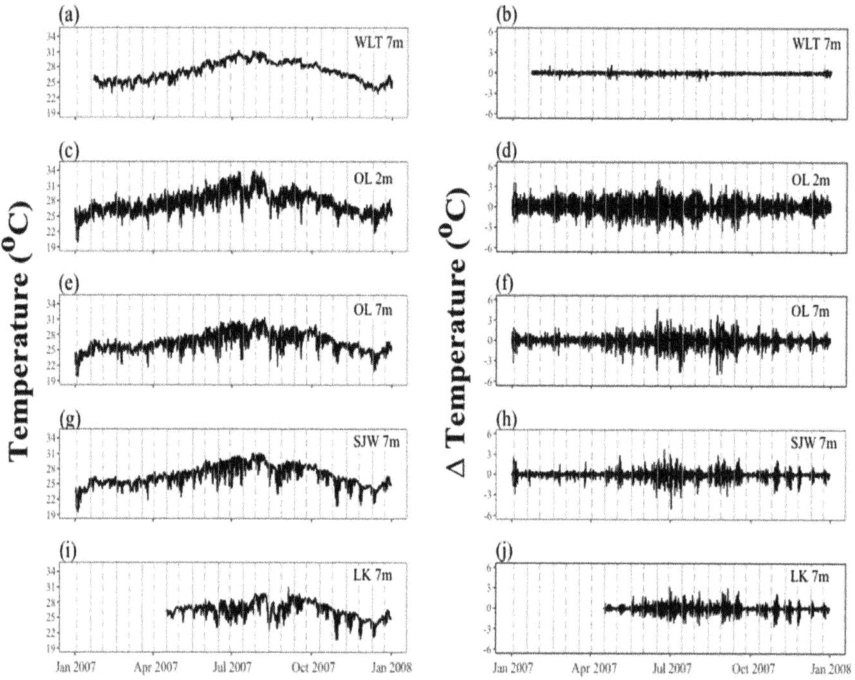

Figure 3. The in situ hourly average seawater temperatures and hourly average temperature differences among Wanlitung (WLT) (7 m deep: **a,b**), Outlet (OL) (2 m: **c,d**; 7 m: **e,f**), Siangjiaowan (SJW) (7 m: **g,h**), and Longkeng (LK) (**i,j**) from 1 January 2007 to 1 January 2008. The vertical dotted and solid lines indicate full moon and new moon, respectively. The seawater temperature was recorded using in situ underwater data loggers (Hobo Pendant® Temperature data logger, Onset Computer Corporation, Bourne, MA, USA). We calculated the hourly average temperature first because temperature was measured at different time interval, such as every 15 min or every 30 min, among sites. Hourly average temperature differences were defined as the difference in hourly average temperatures between any given hour and the hour next. We used hourly instead of daily average temperature difference in order to reveal the temperature difference caused by TIU instead of day cycle.

Table 2. Characteristics of the in-situ temperature record at the five sites from 1 January 2007 to 1 January 2008.

Subregions	Site	Depth (m)	Yearly Mean Temperature	Max SST	Min SST	Max. Variability within 2 Hours
West Hengchun Peninsula	Wanlitung	7	27.090	31.33 (July)	22.77 (Dec.)	3.15 (July)
West Nawan	Outlet	2	27.580	34.02 (July)	19.91 (Jan.)	4.08 (July)
West Nawan	Outlet	7	26.620	31.36 (Aug.)	18.70 (Jan.)	4.90 (Aug.)
East Nawan	Siangjiaowan	7	26.580	31.09 (July)	17.82 (Jan.)	4.99 (July)
East Hengchun Peninsula	Longkeng	7	26.460	30.93 (Sep.)	20.63 (Oct.)	3.13 (June)

This section can be summarized as; (1) There is a presence of strong tidally induced upwelling in Nanwan and (2) Significant temperature drops caused by upwelling could possibly play a major role in shaping coral community structure and long-term resilience in different sub-regions of KNP

2.1. Spatial and Temporal Variability of Coral Communities in Responding to Large-Scale Disturbances

The coral reefs in KNP have been explored since the latter half of Japanese rule, when efforts began to describe coral species and produce a checklist [66–73]. Efforts to describe reef geomorphology and coral communities began in the 1970s, when KNP was established and the third nuclear power plant was constructed [73–80]. These studies focused on describing the reef morphology, coral and fish diversity, and ecology of planktons and fishes. The coral reefs in KNP are separated by sand channels of various widths and vary in their coral fauna among different locations summarised in [79]. There is no significant correlation between species diversity (Shannon-Wiener's index, H') per 10 m line intercept transect and depth, except at the northern tip of Nanwan, where the species diversity decreases with depth [79,80]. It was concluded that KNP contains two types of coral communities, one mainly composed of scleractinian corals with a few alcyonaceans (less than 6% in total coverage) in the wave-protected areas, and the other dominated by alcyonaceans in wave-exposed areas (total coverage > 50%) [80].

Local tourism significantly increased in the 1990s, and long-term ecological research (LTER) was introduced to monitor the impact of local human disturbances—such as overfishing, habitat destruction, and pollution—on KNP coral reefs [59,60,62]. Observations have shown that variations in large-scale physical disturbances, such as typhoons, in different sub-regions had more of an impact on environmental and biological processes than human disturbance, and resulted in spatial heterogeneity of coral communities in KNP [78]. Analysis showed that typhoons in Taiwan took four types of routes between 1911 and 2018: 26.3% traveled north-westward (Type I), 12.3% traveled northward along the east coast (Type II), 10.26% traveled south-eastward (Type III), and 6.84% traveled northward along the west coast (Type IV) (Figure 4). Interestingly, historic records suggest that Type I typhoons are the major contributor of large-scale disturbances that shape coral communities in KNP (Table 3).

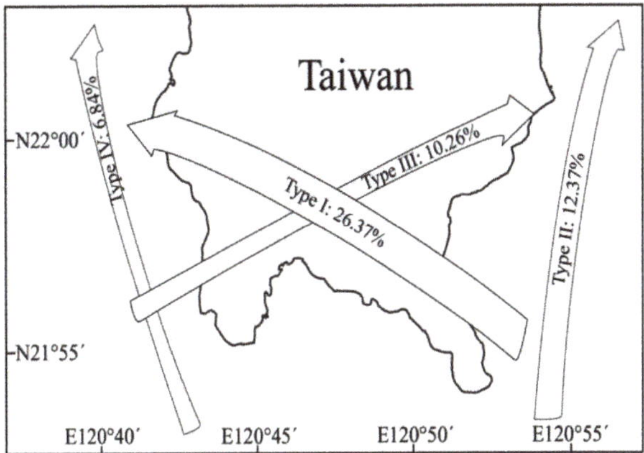

Figure 4. Common trajectories that typhoons took through southern Taiwan between 1911 and 2018. Type I and III indicate a typhoon running north-westwards and north-eastwards, respectively. The other two indicate a typhoon running northwards along the east (type II) and west coasts (type IV). The data were collected from the Typhoon Database, Central Weather Bureau, Taiwan.

Table 3. Summary of the historical disturbances, typhoons, coral bleaching events, and biological outbreaks in Kenting National Park since 1986. Type I and IV of typhoons referred the direction of typhoon used in Figure 4. Typhoon categories are inside the brackets after typhoon names.

Category	Name	Year	Note
Typhoon	Peggy (5)	1986	Type I
Typhoon	Gerald (4)	1987	Type I
Typhoon	Lynn (5)	1987	Type I
Typhoon	Herb (5)	1996	Type I
Typhoon	Chanchu (4)	2006	Type IV
Typhoon	Morakot (2)	2009	Type I
Typhoon	Nanmadol (5)	2011	Type I
Typhoon	Usagi (4)	2013	Type I
Typhoon	Soudelor (5)	2015	Type I
Typhoon	Meranti (5)	2016	Type I
Typhoon	Megi (4)	2016	Type I
Temperature anomaly	Bleaching	1998	Nearly all the colonies on the reefs shallower than 5m in depth in Outlet were bleached.
Temperature anomaly	Bleaching	2002	Minor, very small scale and local bleaching event were recorded in Wanlitung, Houbihu, and Siangjiaowan
Temperature anomaly	Bleaching	2007	50% In Outlet and up to 25% on the West coast of Hengchun Peninsula and Nanwan
Temperature anomaly	Bleaching	2010	Minor scale on the hallow reef of the NPP OL
Temperature anomaly	Bleaching	2014	There were around 30% of the corals bleaching in KNP except in the Outlet that 50% and 20% of the corals were bleached on the shallow (shallower than 5m in depth) and deep (10m in depth) reefs.
Temperature anomaly	Bleaching	2016	Minor scale from Outlet to Nanwan beach
Temperature anomaly	Bleaching	2017	Minor scale on the West coast of Hengchun Peninsula
Ship grounding	Amorgos	2001	Limited on the East coast of Hengchun Penunsula, in particular Longkeng
Ship grounding	Colombo Queen	2009	East coast of Hengchun Peninsula
Ship grounding	WO-BUDMO	2009	West coast of Hengchun Peninsula
Biological outbreak	Sea anemone *Condylactis* sp.	Late 1996–2008	Limited in the shallow area of Tiaoshi
Biological outbreak	Green alga *Codium edule*	Late 1996–2002	Limited in the shallow area of Tiaoshi with significantly seasonal variation

LTER data of 5 to 10 m depths showed a declining trend in mean living coral cover (LCC) in KNP, from 48.56% in 1986 to 29.33% in 2018 (Figure 5a), mainly due to the synergistic effects from multiple disturbances (Table 3). Three time-intervals—1986–2000, 2000–2006, 2006–present—shed light on coral community dynamics. First, although no data is available between 1987 and 1997, typhoons in 1986, 1987, and 1996 and the 1998 mass bleaching event were proposed to account for LCC decreasing to 36% by 1999 [81,82], and as a result of local pollution and habitat destruction, was followed by outbreaks of the macroalgae *Codium* spp. [83] and sea anemone *Condylactis* sp. in Taioshi [84] and see references in Table 1 [62,79–82,85–101]. Second, between 2000 and 2006, the LCC returned to a level similar to that of 1986 (> 45%) in 2003, 2004, and 2005 (Figure 5a) due to a lack of major typhoons or bleaching (except a minor one in 2002) between 1999 and 2005 [62,101].

Interestingly, dominant species that were recovered and found to contribute to the increasing LCC during this period were significantly different from those in 1986; for example, in Wanlitung, a LTER site on the west coast of Hengchun Peninsula, the LCC was composed of *Montipora*, *Heliopora*, and *Poritidae* during this period, whereas *Acropora* was the dominant coral genus in 1986 [62,101]. Third, intense disturbances by typhoons (Table 3) caused LCC in KNP to decline in 2006 and stay low until 2016 (Figure 5a). For example, Typhoon Morakot in 2009, the deadliest one in the recorded history (although recorded as category 2), stayed on top of Taiwan for 2 days, causing flooding and big waves that brought the LCC down to 21.07% in 2010 [102,103].

Although the LCC recovered by 2016 to 43.86%, typhoons Meranti and Megi (ranked category 5 and 4, respectively, Table 3), directly hit KNP in September 2016 and combined with minor coral bleaching in 2017, again caused a dramatic decline of LCC down to 29.33% in 2018 [103].

Spatial variability in the response of coral reefs to disturbances was observed among sub-regions in KNP (Figure 5b–e) by analysing the coral cover using the method described in [104]. The coral reef at the west coast of Hengchun Peninsula (Figure 5b) and east coast of Nanwan (Figure 5e) showed a similar trend of LCC dynamics of the overall KNP, but the latter showed no LCC decline between 1986 and 2000. The east coast of Hengchun Peninsula showed an increasing trend at the time interval between 2000 and 2006, declined due to the impact of typhoon Morakot in 2009, and reached its highest LCC of 44.77% in 2012 before declining again between 2012 and 2016 and, finally recovering by 2018 (Figure 5d). The west coast of Nanwan has maintained a higher LCC (> 40%) than other subregions in the last 36 years (Figure 5c) due to the lack of direct impact from typhoons.

Although typhoons have played a notable role of causing the decline in LCC in KNP, it seems they also have benefited the coral community in KNP. The positive effect of typhoons in terms of cooling of the sea surface temperature and breaking down accumulated heat stress in summer by mixing the heated surface water with cooler water from deeper areas [105,106] has resulted in reduced or no coral bleaching. The best example of this in the KNP is NPP Outlet (OL) located at the west coast of Nanwan. This site is most protected from storm surges and at the same time most exposed to the thermal stress caused by the nuclear power plant discharge. In September 2009, while the storm surges of typhoon Morakot did not cause a significant damage on the LCC of the reefs in NPP Outlet (OL) (Figure 5c), compared to the reefs on the east coast of Nanwan (Figure 5e), it did reduce the temperature of the constantly heated sea surface water at the shallow part of the OL by > 6 °C (Figure 6). The cooling effect created an environmental condition of temperature equal to winter and lasted for 3 days after the typhoon passed.

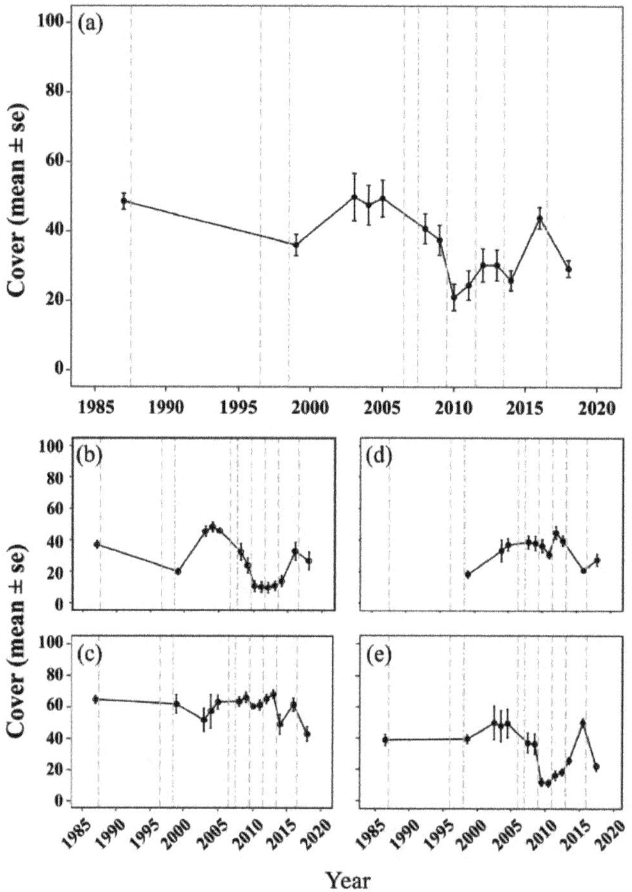

Figure 5. Spatial and temporal long-term trend of average living total coral cover (LCC, per transect ± standard error) in KNP from 1985 to the present. (**a**) the average LCC in the Kenting National Park scale, (**b**) the west, and (**d**) the east coast of Hengchun Peninsula; (**c**) the west, and (**e**) the east coast of Nanwan. The wide and narrow dashed lines indicate typhoons and bleaching events in KNP. The data were collected from the dataset listed in Table 1. The detailed survey methods of the data used in this figure are listed in Table 1. In order to compare the LCC at similar depth, only the transects laid out between 5 and 10m in depth at each site from 1986 to 1999 (Dai, C.-F. 1988; Dai, C.-F. etc. 1998, 1999) were used to generate the figure, combined with the data collected after 2003, including data from a PhD thesis (Dai, C.-F. 1988), one article published in the proceeding of the 6th International Coral Reef Symposium Dai, C.-F. (1988) and two local journal articles written in Mandarin with English abstract (Dai, C.-F. etc. 1998, 1999). The data of LCC after 2003 were collected by the co-first author C.-Y.K. and the corresponding author C.A.C. Different parts of this dataset have been published as a Master's thesis (Kuo, C.-Y. 2007), in local project reports (Kuo, C.-Y. etc. 2012; Shao, K.-T. etc. 2002; Fang, L.-S. etc. 2003–2006; Wang, W.-H. etc. 2007; Ho, M.-J. etc. 2016; Ho, P.-H. etc. 2008–2011; Chen, C.A. etc. 2012–2014, 2016, 2018; Kuo, C.-Y. etc. 2007 and journal articles (Kuo, C.-Y. etc. 2011, 2012).

The coral communities in the reef adjacent to the nuclear power plant outlet (OL) are protected by storm surges and cooled down by TIU and typhoons, however, the local rising and variable sea surface temperatures [107–110] has resulted in shifting dominant coral species (Figure 7) due to multiple coral bleaching events over time (Table 3). Also at NPP OL, the warm discharged water is trapped in the shallow waters (up to 4 m deep) and flows southwestward in Nanwan because of a near-shore current and tides [111], resulting in a 2.0–3.0 °C higher summer average seawater temperature than at other coral reef sites in Kenting [107,112,113]. Comparing the living coral assemblage in 1986, 1995, 2010, and 2019 in shallow water (3 m) at OL, there was a sharp change in coral genus composition in 1995 (Figure 7a). While *Acropora* dominated in 1986 (31.58% of relative LCC) and 1995 (59.42%), *Galaxea* replaced it and became dominant in 2010 (31.64%) and 2019 (21.11%). *Montipora* remained relatively constant throughout the monitoring period, and *Seriatopora* and *Stylophora* were completely absent at 3 m at OL after 2010 [107]. In addition, coral genus composition remained similar in 1986, 1995, 2010, and 2019 in deep water (7 m), although their relative abundance fluctuated through time (Figure 7b).

However, the response of coral communities to typhoons are varied and has resulted in spatial variation in long term changes of LCC among subregions in KNP (Figure 5). In the nuclear power plant outlet (OL), the reef most protected from typhoons, local, small scale variation of temperature has caused multiple bleaching events and resulted in the loss of temperature sensitive taxa.

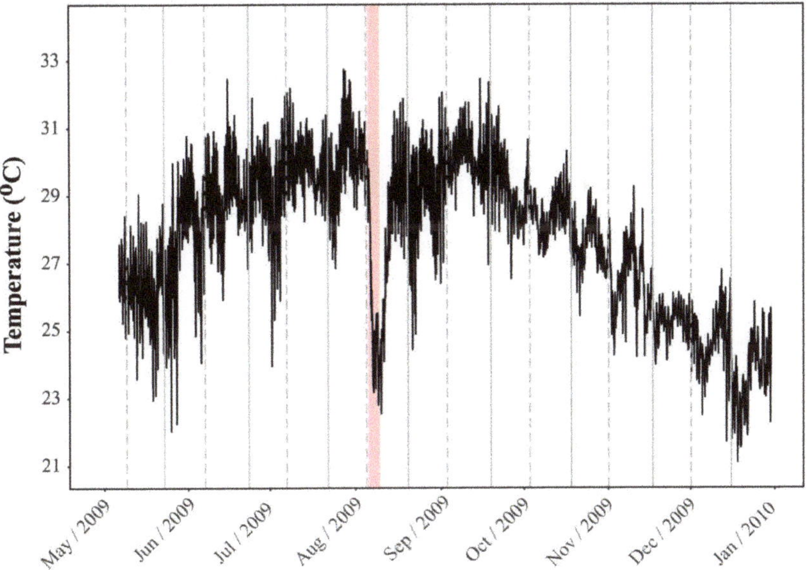

Figure 6. The in situ hourly average seawater temperatures at 2 m depth in Outlet (OL) from 6 June 2009 to 31 December 2009. The vertical dotted and solid lines indicate full moon and new moon, respectively. The pink area indicates the period, from 5th August. 20:30 to 10th August 5:30 2009, the sea warning for typhoon Morakot, the deadliest one in the history of Taiwan, was issued by the Central Weather Bureau.

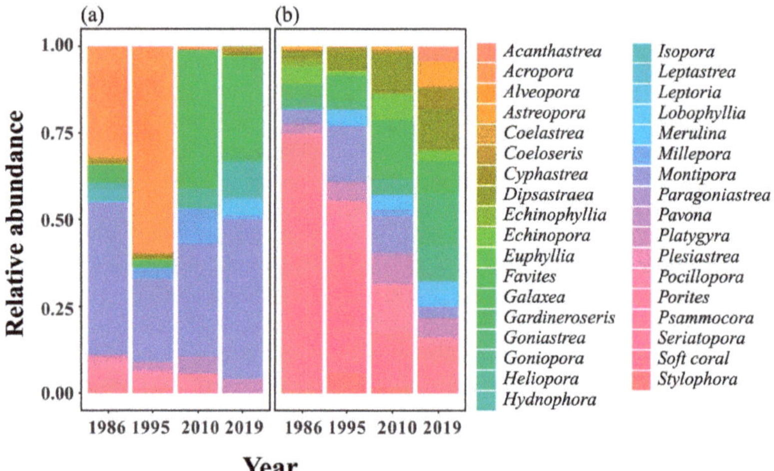

Figure 7. The relative abundance of coral genera at (**a**) 3 m (shallow) and (**b**) 7 m (deep) in NPP OL in 1986, 1995, 2010, and 2019. This figure was redrawn using data in (Keshavmurthy, S. etc. 2014) and combined with a survey conducted in 2019.

This section can be summarized as; (1) The effect of typhoons and bleaching has made notable impact on environmental and biological processes resulting in spatial heterogeneity of coral communities and (2) Response of coral communities to typhoons are varied and has resulted in spatial variation in long term changes of LCC among subregions in KNP. In the nuclear power plant outlet (OL), the reef most protected from typhoons, local, small scale variation of temperature has caused multiple bleaching events and resulted in the loss of temperature sensitive taxa.

2.2. Symbiont Community Dynamics over Space and Time in KNP

Coral-associated Symbiodiniaceae from different locations in KNP have been analysed since 1997 [107–110,114–116]. Our studies on Symbiodiniaceae diversity in KNP over the past 24 years demonstrate fine-scale, micro-geographic, temporal, and species-specific associations among genera in addition to/other than coral-associated *Cladocopium* spp. In particular, the ability of corals to associate with multiple Symbiodiniaceae genera and change (i.e., shuffle) between stress-resistant and stress-sensitive genera/species depending on environmental conditions [46,47,117–122] is a critical requirement of their resilience towards stress. At the community level, *Cladocopium* spp. were the dominant Symbiodiniaceae associated with corals in KNP (previously Clade C), as they are elsewhere in the Pacific and South China Sea [123–125]. No matter the molecular technique applied to phylotype Symbiodiniaceae, the results consistently shows that *Cladocopium* sp. is dominant and co-occurs with *Durusdinium* sp. (previously Clade D).

Fine-scale techniques such as ITS2-DGGE, used from 2009 onwards, revealed a fine-scale variation in associated species within each genus; for example, *Cladocopium* C1, C3 and *Durusdinium trenchii* (previously D1a) have all been shown to be dominant. However, recent use of up-to-date NGS amplicon sequencing has revealed more diversity within genera *Cladocopium* and *Durusdinium* (Table 4). In addition, a study published in 2014 [107] showed depth and species-related differences in the coral-associated Symbiodiniaceae in KNP. Samples collected from 16 genera from eight locations and two depths in KNP revealed some interesting trends (Figure 8). *Cladocopium* spp. were more dominant in deeper than shallow water, especially in the corals occurring near OL and in Nanwan.

Table 4. Coral-Symbiodiniaceae associations in Kenting National Park through time analyzed using more sensitive over time. Until 2010 all data are shown by old taxonomy used for Symbiodiniaceae. C = *Cladocopium* spp. (previously clade C), D = *Durusdinium* spp. (previously clade D).

Year	Host Species (Family/Genus)	Symbiodiniaceae Clade/Type/Genera/Species	Study Sites in KNP	Genetic Method for ID	Reference
1997–2001	*Acropora*	C3, C1, D1, D2,		srDNA- RFLP	Chen, C.A. etc. 2005
	Montipora	C1			
	Pocilloporidae	C1, C2			
	Euphyllidae	C1, D1			
	Poritidae	C1			
	Siderastreidae	C1			
	Agariciidae	C1			
	Oculinidae	C3			
	Merulinidae	C1			
	Faviidae	C1			
2000–2001	*Isopora palifera*	C, D	Tantzei Bay	srDNA- RFLP	Hsu, C.-M. etc. 2012
2006–2009	*Isopora palifera*	C3, D1a	Tantzei Bay, Maobitou, Siatanzai, NPP-OL, Shiaowan, Siangjiaowan Longkeng	srDNA- RFLP ITS2-DGGE	
2009	*Platygyra verweyi*	C3, D1a	Leidashih, Siatanzai Maobitou, NPP-OL Wanlitung, Hungchai NPP-IL, Tiaoshi, Tantzei Bay, Longkeng	ITS2-DGGE ITS1-qPCR	Keshavmurthy, S. etc. 2012
2009–2010	*Acanthastrea*	C1, D1a	Houbihu, NPP-OL	srDNA- RFLP	Keshavmurthy, S. etc. 2014
	Acropora	C21a, C3, D1a	Siangjiaowan, NPP-IL	ITS2-DGGE	
	Cyphastrea	C3, D1a	Wanlitung, Tiaoshi	ITS1-qPCR	
	Favia	C3, D1a	Tantzei Bay, Longkeng		
	Favites	C3, D1a			
	Galaxea	C1, D1a			
	Goniastrea	C1, D1a			
	Isopora	C3, D1a			
	Leptastrea	D1a			
	Leptoria	C1, D1a			
	Montastrea	C1, C3, D1a			
	Montipora	C15, D1a			
	Pavona	C1, D1a			
	Platygyra	C3, D1a			
	Pocillopora	C3, D1a			
	Porites	C15, D1a			
	Seriatopora	C1			
	Stylophora	C1			

Table 4. *Cont.*

Year	Host Species (Family/Genus)	Symbiodiniaceae Clade/Type/Genera/Species	Study Sites in KNP	Genetic Method for ID	Reference
2016–2017	*Leptoria phyrgia*	*Durusdinium glynii*	Wanlitung	ITS2-DGGE	Carballo-Bolaños, R. etc. 2019
		Durusdinium trenchii *Cladocopium* C3w *Cladocopium* C21a *Cladocopium* sp.	NPP-OL	ITS1-qPCR	
2019	*Leptoria phyrgia*	*Durusdinium glynii*	Wanlitung	ITS2 Amplicon	Huang, Y-Y. etc. 2019
		Durusdinium trenchii *Durusdinium* D1.6, D17, D2, D5, D6 *Cladocopicum* C116, C15.7, C21a, C2r, C3.1 C3.8, C3b, C3d, C3e, C3s, C50	NPP-OL		

RFLP = Restriction Fragment Length Polymorphism, DGGE = Denaturing Gradient Gel Electrophoresis, qPCR = Quantitative Real Time Polymerase Chain Reaction.

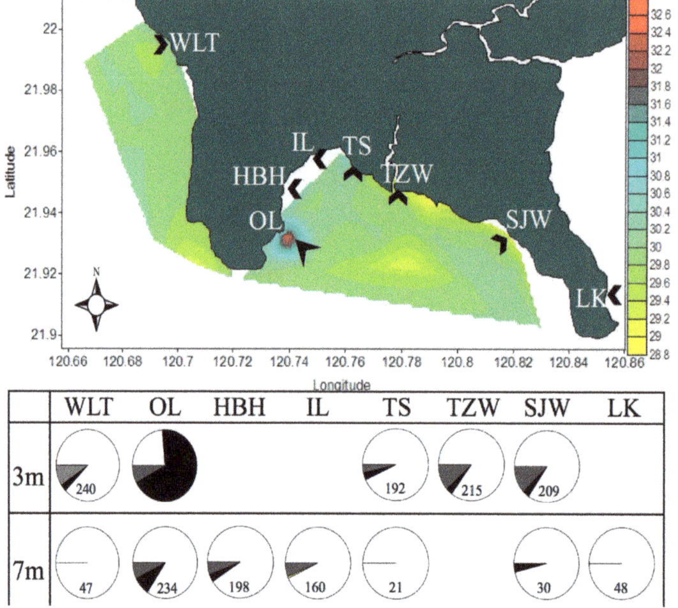

Figure 8. Symbiont diversity associated with corals at different locations—Wanlitung (WLT), NPP outlet (OL), Houbihu (HBH), NPP Inlet (IL), Taioshi (TS), Tanziwan (TZW), Siangjiaowan (SJW), and Longkeng (LK)—and from the reef near the nuclear power plant outlet (OL) in Kenting National Park. Samples were collected in 2009 and 2010 from 3 m and 7 m depths. Pie-Charts were drawn using data from previous publications (Keshavmurthy, S. etc. 2014). Analysis of samples collected was done using srDNA-RFLP, ITS2-DGGE and ITS1-qPCR (see Table 4). White = *Cladocopium* sp. (previously clade C), Black = *Durusdinium* sp. (previously clade D), and Grey = co-occurrence of *Cladocopium* sp. and *Durusdinium* sp. The values inside the pie-charts are total sample numbers for each location.

For example, *Isopora palifera* samples in the Tanziwan (TZW) population were dominated by *Durusdinium* spp. in 2001 (three years after the 1998 mass coral bleaching event), *Cladocopium* spp. in

2005, co-dominant by two Symbiodiniaceae genera in 2009, and returned to *Durusdinium* dominance in 2015 (Figure 9). This is concordant with previous studies that show that the occurrence of multiple Symbiodiniaceae genera at low concentrations densities might lead to either shuffling or switching to beneficial Symbiodiniaceae genera over time [121,122,126]; in some cases, the coral host may revert back to its original composition of either a single dominant Symbiodiniaceae species or multiple dominant species/genera [108,122]. In addition to temporal changes, our study also indicated a very efficient spatial difference in Symbiodiniaceae associated with *I. palifera*. Samples collected from different locations in KNP showed [108] that corals at a site next to the OL were exclusively associated with or dominated by *D. trenchii/glynii,* and those at sites away from the OL were associated with *Cladocopium* C3 (Figure 9). Such a micro-geographic difference in Symbiodiniaceae composition is due to the presence of the OL, which has released hot water onto to the coral reefs for over 35 years. Hence, corals near the OL have adapted/acclimatized to associate with *Durusdinium* spp. that are generally stress- as well as temperature-tolerant. Similar spatial differences in the association with Symbiodiniaceae was found in the coral *Platygyra verweyi* and *Leptoria phygria* [108,110]. Nonetheless, there are also cases, irrespective of environmental perturbations, in which the host maintains stable symbiosis with a particular Symbiodiniaceae genus [123,127,128].

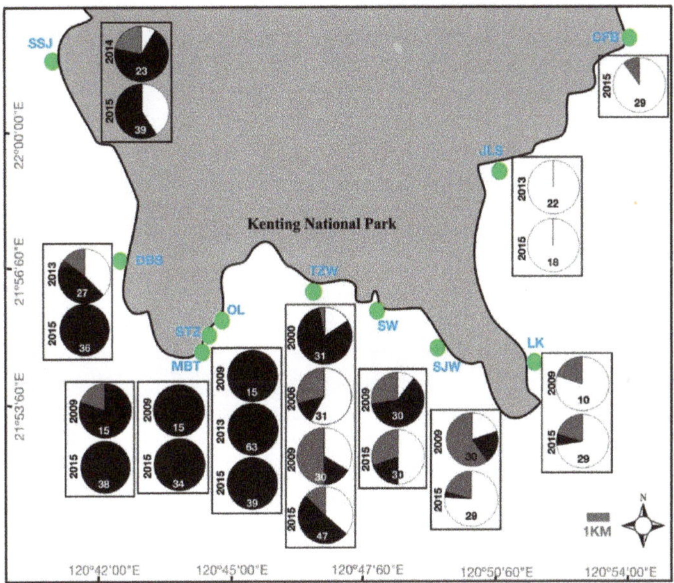

Figure 9. Spatial and temporal variation in *Cloadopoium* sp. and *Durusdinium* sp. data in the coral *Isopora palifera* from different locations—Siashuijue (SSJ), Dingbaisha (DBS), Maobitou (MBT), Siatanzai (STZ), nuclear power plant outlet (OL), Tanziwan (TZW), Shiaowan (SW), Siangjiaowan (SJW), Longkeng (LK), Jialeshuei (JLS), and Chufengbi (CFB) in Kenting National Park. The data were collected from 2000 to 2015 are adjacent to the pie-charts. Pie-Charts were drawn using data from previous publications (Hsu, C.-M etc. 2012; Fong, W.-L. 2016). Analysis of samples collected in 2000 was done using srDNA-RFLP and all the samples obtained between 2006–2015 were analysed by srDNA-RFLP and ITS2-DGGE (see Table 3) White = *Cladocopium* sp. (previously clade C), Black = *Durusdinium* sp. (previously clade D), and Grey = co-occuurence of *Cladocopium* sp. and *Durusdinium* sp. The values inside the pie-charts are total sample numbers for every location.

This section can be summarized as; (1) Dominant Symbiodiniaceae associated with corals in the KNP is Cladocopium sp. However, majority of coral species in the shallows of NPP OL are associated with Durusdinium sp. Shuffling of Symbiodiniaceae in corals is almost non-existent.

3. Discussion and Conclusions

There is a serious concern that coral reefs will almost entirely disappear by 2050 if average ocean temperatures increase by 2 °C, with just 10–30% of existing reefs surviving if ocean temperatures increase by 1.5 °C [19]. The "mission-impossible" goal is to drastically reduce CO_2 emissions to net zero and maintain an only 1.5 °C temperature increase; this might lead to 10% of current reefs surviving after 2050. Some global efforts, such as the 50 Reefs Initiative, have used Modern Portfolio Theory (MPT) to identify coral reef locations that represent imperative conservation investments and to ensure their survival; the goal of these efforts is to prepare these areas for repopulation once the climate has been stabilized [129]. If 10%–30% of existing reefs will indeed survive if ocean temperatures increase by only 1.5 °C, an important question is what to do if a reef does not pass the MPT criteria, but does possess certain localized environmental settings, reef topologies, and coral species that do or could resist the impacts of climate change. Herein we argue that the coral reef in KNP could have this great potential to resist the impacts of climate change and deserves novel conservation efforts to ensure its contribution to the coral reef resilience not only in Taiwan but also in the West Pacific.

First, the KNP coral reefs probably receive a constant supply of coral larvae from the south to replenish the reefs after disturbances, and also serve as a "stepping stone" to connect to the reefs or coral communities at high latitudes with currents such as the Kuroshio Current (KC) and the South China Sea Surface Current (SCSSC). Southern Taiwan is bordered by Luzon Island (the Philippines), the north boundary of the "Coral Triangle" (CT). It is argued that KNP has a relatively high scleractinian and reef-associated species diversity because of its connection to the CT [52–54]. Preliminary studies on the genetic connectivity of coral reef fishes in the West Pacific and South China Sea support this scenario [130,131]. Further research on the genetic connectivity of scleractinian corals using high-resolution molecular markers could help elucidate the resilience role of the KNP reef in the region.

Second, many typhoons in the north-western Pacific pass through Taiwan; thus, typhoons might play both negative and positive roles in shaping the coral community structure at different sub-regions in KNP. Of the four types of typhoons recorded (Figure 4), type I (Table 3)—created by a southeast-northwest vortex—contributes most to the temporal dynamics and spatial heterogeneity of coral communities in different sub-regions of KNP (Figure 5). While there is always some mechanical damage, KNP coral reefs can also benefit from typhoons during the warm summer months. As ocean surface waters become warmer during the summer, corals often experience thermal stress. Typhoons can relieve thermal stress by (1) absorbing energy from surface waters through the transfer of latent heat; (2) inducing local upwelling, bringing deeper, cooler water to the surface; and (3) creating clouds of typhoons to shade the ocean surface from solar heating, allowing the water to cool and reducing light stress. Although the projected impacts of climate change on typhoons remain debatable, monitoring the physical damages caused by typhoons and their joint effect with thermal-induced coral bleaching will be crucial for us to develop a management plan to improve coral reef resilience in KNP.

Third, upwelling has been proposed to be a cooling mechanism to protect coral reefs against bleaching by reducing seawater temperatures or creating fluctuating thermal environments that induce corals to build thermal resistance over time [132–135]. However, it has also been suggested that upwelling areas do not always guarantee refuge for coral reefs in a warming ocean unless the thermal threat and upwelling coincide [136]. Some large-scale seasonal upwelling with cold, nutrient-rich, and naturally acidic subsurface water—such as the upwelling in Gulf of Panama and Papagayo upwelling of Costa Rica, in the tropical eastern Pacific—indeed hinders the development of coral reefs [137–139]. In KNP, upwelling is small-scale, localized, and induced by tides flowing from east to west Nanwan that create a temperature difference within the bay of Nanwan and two sides of Hengchun Peninsula, where the water is cooler in the east and warmer in the west (Table 2).

In addition, the upwelling helps reduce the thermal stress, particularly in the reef adjacent to the nuclear power plant (OL), by creating significant temperature drops during spring tides in the summer (Table 2) and fluctuating thermal environments that induce corals to build thermal resistance (Figures 8 and 9). These positive effects ensure that KNP remains a refuge for coral reefs to survive in a warming ocean. Monitoring whether the tide-induced upwelling will be enhanced or hindered by the rising background seawater temperature in the region should be considered as a research priority in KNP.

Fourth, Symbiodiniaceae play a crucial role in bleaching tolerance. Many species or genera of Symbiodiniaceae have been identified [44], and different genera display varying thermal, and therefore bleaching, resistances. It has been suggested that, by associating with or shuffling the symbiont community towards making thermal-tolerant Symbiodiniaceae, such as *Durusdinium* spp, dominant, corals can increase their thermal tolerance by 1.0 °C–1.5 °C [118,119]. Our long-term monitoring of symbiont community diversity shows that corals constantly exposed to warming and fluctuating thermal environments (OL) or constantly higher seawater temperatures (west coast of Hengchun Peninsula) have a dominance of *Durusdinium* spp, whereas the same species located on the cooler east coast of Hengchun Peninsula are increasingly associated with *Cladocopium* spp. (Figure 8). This subregional difference in symbiont community is concordant with the influence of TIU that pumps cooler water from Eluanbi (east cape), protruding towards Maobitou (east cape), creating two temperature drops per tidal cycle in western and central Nanwan and one drop in the eastern part, but no having impact on the west coast of the peninsula. The TIU affects the seawater temperature in KNP and not only drives sub-regional variability in symbiont communities, but also provides the signal for corals to shuffle their symbionts in response to seasonally fluctuating seawater temperature; this is not, however, true for corals in shallow water (< 3 m) of the reef adjacent to the nuclear power plant OL, which are associated dominantly with *Durusdinium* spp. and do not show sign of shuffling [107–109]. In reciprocal transplantation experiments (RTE), corals from WLT to OL did not survive under a prolonged seawater temperature anomaly, even though they showed signs of shuffling from *Cladocopium* to *Durusdinium* dominance [22]. These results imply that corals in the shallow water of the OL reef already live at the ceiling of thermal tolerance, and future climate change trends might be untenable for those corals [110].

Despite these environmental, ecological, and biological characteristics, adaptive management strategies such as implementing sewage treatment systems, banning the serving of herbivorous fishes in restaurants, and promoting eco-friendly tourism and public awareness in recent years has aided in the resilience of coral reefs in KNP. Present and future adaptive management in accordance with the framework of resilience-based management [39] might help sustain coral reef resilience in Kenting National under the impacts of climate change.

This section can be summarized as; Coral reefs in a small geographical range with unique environmental settings and ecological characteristics, such as the KNP reef, are resilient to bleaching and deserve novel conservation efforts. Thus, conservation efforts that use resilience-based management programs to reduce local stresses and meet the challenge of climate change is urgently needed.

Author Contributions: Conceptualization, C.A.C., J.-T.W., P.-J.M., S.K. and C.-Y.K.; Methodology, C.A.C., S.K., C.-Y.K., R.C-B., Y.-Y.H.; Formal analysis, C.-Y.K., S.K., R.C.-B., Y.-Y.H.; Investigation, C.A.C., C.-Y.K., S.K.; Resources, C.A.C., J.-T.W., P.-J.M.; Data curation, S.K., C.-Y.K.; Writing—original draft preparation, S.K., C.A.C., C.-Y.K.; Writing—review and editing, S.K., C.A.C., C.-Y.K.; Funding acquisition, C.A.C., J.-T.W., P.-J.M.

Funding: This research was funded by Academia Sinica for Life Science Research (no. 4010) and Academia Sinica Thematic Grant- Green Island (no. AS-TP-108-LM14) to CAC.

Acknowledgments: Many thanks LM Chou and D Huang, the subject editors of this coral reef resilience special issue, for inviting us to contribute to this long-term ecological research (LTER) data project on Kenting National Park, Taiwan. We thank all the students, research assistants, and postdoctoral fellows over the past two decades that helped collect data for this LTER. S.K. and C.-Y.K. are supported by Academia Sinica postdoctoral fellowships. This work was funded by the Ministry of Science and Technology, Kenting National Park, and Taiwan Power Company to P.-J.M., J.-T.W., and C.A.C. We would like to thank Noah Last of Third Draft Editing for his English language editing.

Conflicts of Interest: The authors declare no conflict of interest and the funders had no role in the design of the study; in the collection, analyses, or interpretation of data; in the writing of the manuscript, or in the decision to publish the results.

References

1. Ateweberhan, M.; Feary, D.A.; Keshavmurthy, S.; Chen, C.A.; Schleyer, M.H.; Sheppard, R.C. Climate change impacts on coral reefs: Synergies with local effects, possibilities for acclimation, and management implications. *Mar. Pollut. Bull.* **2013**, *74*, 526–539. [CrossRef] [PubMed]
2. Hughes, T.P.; Baird, A.H.; Bellwood, D.R.; Card, M.; Connolly, S.R.; Folke, C.; Grosberg, R.; Hoegh-Guldberg, O.; Jackson, J.B.; Kleypas, J.; et al. Climate Change, Human Impacts, and the Resilience of Coral Reefs. *Science* **2013**, *301*, 929–933. [CrossRef] [PubMed]
3. Hoegh-Guldberg, O.; Mumby, P.J.; Hooten, A.J.; Steneck, R.S.; Greenfield, P.; Gomez, E.; Harvell, C.D.; Sale, P.F.; Edwards, A.J.; Caldeira, K.; et al. Coral reefs under rapid climate change and ocean acidification. *Science* **2007**, *318*, 1737–1742. [CrossRef] [PubMed]
4. National Academies of Sciences, Engineering, and Medicine. *A Research Review of Interventions to Increase the Persistence and Resilience of Coral Reefs*; The National Academies Press: Washington, DC, USA, 2018; 258p.
5. Brown, B.E. Coral bleaching: Causes and consequences. *Coral Reefs* **1997**, *16*, S129–S138. [CrossRef]
6. Hoegh-Guldberg, O. Climate change, coral bleaching and the future of the world's coral reefs. *Mar. Freshw. Res.* **1999**, *50*, 839–866. [CrossRef]
7. Hughes, T.P.; Kerry, J.T.; Álvarez-Noriega, M.; Álvarez-Romero, J.G.; Anderson, K.D.; Baird, A.H.; Babcock, R.C.; Beger, M.; Bellwood, D.R.; Berkelmans, R.; et al. Global warming and recurrent mass bleaching of corals. *Nature* **2017**, *543*, 373–377. [CrossRef]
8. Hughes, T.P.; Anderson, K.D.; Connolly, S.R.; Heron, S.F.; Kerry, J.T.; Lough, J.M.; Baird, A.H.; Baum, J.K.; Berumen, M.L.; Bridge, T.C.; et al. Spatial and temporal patterns of mass bleaching of corals in the Anthropocene. *Science* **2018**, *359*, 80–83. [CrossRef]
9. Gilmour, J.P.; Cook, K.L.; Ryan, N.M.; Puotinen, M.L.; Green, R.H.; Shedrawi, G.; Hobbs, J.-P.A.; Thomson, D.P.; Babcock, R.C.; Buckee, J.; et al. The state of Western Australia's coral Reefs. *Coral Reefs* **2019**, *38*, 651–667. [CrossRef]
10. Goyen, S.; Camp, E.F.; Fujise, L.; Lloyd, A.; Nitschke, M.R.; LaJeunensse, T.; Kahlke, T.; Ralph, P.J.; Suggett, D. Mass coral bleaching of *P. verispora* in Sydney Harbour driven by the 2015–2016 heatwave. *Coral Reefs* **2019**, *38*, 815–830. [CrossRef]
11. Head, C.E.I.; Bayley, D.T.I.; Rowlands, G.; Roche, R.C.; Tickler, D.M.; Rogers, A.D.; Koldewey, H.; Turner, J.R.; Andradi-Brown, D.A. Coral bleaching impacts from back-to-back 2015–2016 thermal anomalies in the remote central Indian Ocean. *Coral Reefs* **2019**, *38*, 605–618. [CrossRef]
12. Montefalcone, M.; Morri, C.; Bianchi, C.N. Long-term change in bioconstruction potential of Maldivian coral reefs following extreme climatic anomalies. *Glob. Chang. Biol.* **2018**, *24*, 5629–5641. [CrossRef] [PubMed]
13. Fox, M.D.; Carter, A.L.; Edwards, C.B.; Takeshita, Y.; Johnson, M.D.; Petrovic, V.; Amir, C.G.; Sala, E.; Sandin, S.A.; Smith, J.E. Limited coral mortality following acute thermal stress and widespread bleaching on Palmyra Atoll, central Pacific. *Coral Reefs* **2019**, *38*, 701–712. [CrossRef]
14. Raymundo, L.J.; Burdick, D.; Hoot, W.C.; Miller, R.M.; Brown, V.; Reynolds, T.; Gault, J.; Idechong, J.; Fifer, J.; Williams, A. Successive bleaching events cause mass coral mortality in Guam, Micronesia. *Coral Reefs* **2019**, *38*, 677–700. [CrossRef]
15. Vargas-Ángel, B.; Huntington, B.; Brainard, R.E.; Venegas, R.; Oliver, T.; Barkley, H.; Cohen, A. El Nino-associated catastrophic coral mortality at Jarvis Island, central Equatorial Pacific. *Coral Reefs* **2019**, *38*, 731–741. [CrossRef]
16. Smith, K.M.; Payton, T.G.; Sims, R.J.; Stroud, C.S.; Jeanes, R.C.; Hyatt, T.B.; Childress, M.J. Impacts of consecutive bleaching events and local algal abundance on transplanted coral colonies in the Florida Keys. *Coral Reefs* **2019**, *38*, 851–861. [CrossRef]
17. Teixeira, C.D.; Leitão, R.L.L.; Ribeiro, F.V.; Moraes, F.C.; Neves, L.M.; Bastos, A.C.; Pereira-Filho, G.H.; Kampel, M.; Salomon, P.S.; Sá, J.A.; et al. Sustained mass coral bleaching (2016–2017) in Brazilian turbid-zone reefs: Taxonomic, cross-shelf and habitat-related trends. *Coral Reefs* **2019**, *38*, 801–813. [CrossRef]
18. Eakin, M.C.; Sweatman, H.P.A.; Brainard, R.E. The 2014–2017 global-scale coral bleaching event: Insights and impacts. *Coral Reefs* **2019**, *38*, 539–545. [CrossRef]

19. Masson-Delmotte, V.; Zhai, P.; Pörtner, H.O.; Roberts, D.; Skea, J.; Shukla, P.R.; Pirani, A.; Moufouma-Okia, W.; Péan, C.; Pidcock, R.; et al. (Eds.) IPCC, 2018: Summary for Policymakers. In *Global Warming of 1.5 °C. An IPCC Special Report on the Impacts of Global Warming of 1.5 °C above Pre-Industrial Levels and Related Global Greenhouse Gas Emission Pathways, in the Context of Strengthening the Global Response to the Threat of Climate Change, Sustainable Development, and Efforts to Eradicate Poverty*; World Meteorological Organization: Geneva, Switzerland, 2018; 32p.

20. Riegl, B.M.; Purkis, S.J.; Al-Cibahy, A.S.; Abdel-Moati, M.A.; Hoegh-Guldberg, O. Present limits to heat-adaptability in corals and population-level responses to climate extremes. *PLoS ONE* **2011**, *6*, e24802. [CrossRef]

21. Pandolfi, J.M.; Connolly, S.R.; Marshall, D.J.; Cohen, A.L. Projecting Coral Reef Futures Under Global Warming and Ocean Acidification. *Science* **2011**, *333*, 418–422. [CrossRef]

22. Kao, K.-W.; Keshavmurthy, S.; Tsao, C.-H.; Wang, J.-T.; Chen, C.A. Repeated and Prolonged Temperature Anomalies Negate Symbiodiniaceae Genera Shuffling in the Coral *Platygyra verweyi* (Scleractinia; Merulinidae). *Zool. Stud.* **2018**, *57*. [CrossRef]

23. Emanuel, K.A. Downscaling CMIP5 climate models shows increased tropical cyclone activity over the 21st century. *Proc. Natl. Acad. Sci. USA* **2013**, *110*, 12219–12224. [CrossRef] [PubMed]

24. Tsuboki, K.; Yoshioka, M.K.; Shinoda, T.; Kato, M.; Kanada, S.; Kitoh, A. Future increase of super typhoon intensity acclimated with climate change. *Geophys. Res. Lett.* **2015**, *42*, 646–652. [CrossRef]

25. Knutson, T.R.; McBride, J.L.; Chan, J.; Emanuel, K.; Holland, G.; Landsea, C.; Held, I.; Kossin, J.P.; Srivastava, A.K.; Sugi, M. Tropical cyclones and climate change. *Nat. Geosci.* **2010**, *3*, 157–163. [CrossRef]

26. Christensen, J.H.; Kanikicharla, K.K.; Marshall, G.; Turner, J. Climate phenomena and their relevance for future regional climate change. In *Climate Change 2013: The Physical Science Basis*; Stocker, T.F., Qin, D., Plattner, G.-K., Tignor, M., Allen, S.K., Boschung, J., Nauels, A., Xia, Y., Bex, V., Midgley, P.M., Eds.; Cambridge University Press: Cambridge, UK, 2013; pp. 1217–1308.

27. Walsh, K.J.; McBride, J.L.; Klotzbach, P.J.; Balachandran, S.; Camargo, S.J.; Holland, G.; Knutson, T.R.; Kossin, J.P.; Lee, T.; Sobel, A.; et al. Tropical cyclones and climate change. *WIREs Clim. Chang.* **2016**, *7*, 65–89. [CrossRef]

28. Dollar, S.J. Wave stress and coral community structure in Hawaii. *Coral Reefs* **1982**, *1*, 71–81. [CrossRef]

29. Cheal, A.J.; MacNeil, M.A.; Emslie, M.J.; Sweatman, H. The threat to coral reefs from more intense cyclones under climate change. *Glob. Chang. Biol.* **2017**, *23*, 1511–1524. [CrossRef]

30. Holling, C.S. Resilience and stability of ecological systems. *Annl. Rev. Ecol. Syst.* **1973**, *4*, 1–23. [CrossRef]

31. Nyström, M.; Folke, C. Spatial resilience of coral reefs. *Ecosystems* **2001**, *4*, 406–417. [CrossRef]

32. Nystrom, M. Redundancy and response diversity of functional groups: Implications for the resilience of coral reefs. *AMBIO* **2006**, *35*, 30–35. [CrossRef]

33. Norström, A.; Nyström, M.; Lokrantz, J.; Folke, C. Alternative states of coral reefs: Beyond coral-macroalgal phase shifts. *Mar. Ecol. Prog. Ser.* **2009**, *376*, 295–306. [CrossRef]

34. Nyström, M.; Norström, A.; Blenckner, T.; de la Torre-Castro, M.; Eklöf, J.S.; Folke, C.; Österblom, H.; Steneck, R.S.; Thyresson, M.; Troell, M. Confronting Feedbacks of Degraded Marine Ecosystems. *Ecosystems* **2012**, *15*, 695–710. [CrossRef]

35. Hughes, T.P.; Graham, N.A.J.; Jackson, J.B.C.; Mumby, P.J.; Steneck, R.S. Rising to the challenge of sustaining coral reef resilience. *Trends Ecol. Evolut.* **2010**, *25*, 633–642. [CrossRef] [PubMed]

36. Montefalcone, M.; Parravicini, V.; Bianchi, C.N. Quantification of cosstal ecosystem resilience. In *Treatise on Estuarine and Coastal Science*; Elsevier: Amsterdam, The Netherlands, 2011; Volume 10, pp. 49–70.

37. Folke, C. Resilience: The emergence of a perspective for social–ecological systems analyses. *Glob. Environ. Chang.* **2006**, *16*, 253–267. [CrossRef]

38. Nystrom, M.; Graham, A.J.; Lokrantz, J.; Norstorm, A.V. Capturing the cornerstones of coral reef resilience: Linking theory to practice. *Coral Reefs* **2008**, *27*, 795–809. [CrossRef]

39. Anthony, K.R.N. Coral Reefs Under Climate Change and Ocean Acidification: Challenges and Opportunities for Management and Policy. *Annu. Rev. Environ. Res.* **2016**, *41*, 59–81. [CrossRef]

40. Barneah, O.; Brickner, I. Multibiont Symbioses in the Coral Reef Ecosystem. In *Symbioses and Stress. Cellular Origin, Life in Extreme Habitats and Astrobiology*; Seckbach, J., Grube, M., Eds.; Springer: Dordrecht, The Netherlands, 2010; Volume 17.

41. Baker, A.C. Flexibility and specificity in coral-algal symbiosis: Diversity, ecology, and biogeography of *Symbiodinium*. *Annu. Rev. Ecol. Evol. Syst.* **2003**, *34*, 661–689. [CrossRef]

42. Lajeunesse, T.C.; Pettay, D.T.; Sampayo, E.E.; Phongsuwan, N.; Brown, B.E.; Obura, D.O.; Hoegh-Guldberg, O.; Fitt, W.K. Long-standing environmental conditions, geographic isolation and host-symbiont specificity influence the relative ecological dominance and genetic diversification of coral endosymbionts in the genus *Symbiodinium*. *J. Biogeogr.* **2010**, *37*, 785–800. [CrossRef]

43. Silverstein, R.N.; Correa, A.M.S.; Baker, A.C. Specificity is rarely absolute in coral-algal symbiosis: Implications for coral response to climate change. *Proc. R. Soc. B Biol. Sci.* **2012**, *279*, 2609–2618. [CrossRef]

44. Lajeunesse, T.C.; Parkinson, J.E.; Gabrielson, P.W.; Jeong, H.J.; Reimer, J.D.; Voolstra, C.R.; Santos, S.R. Systematic Revision of Symbiodiniaceae Highlights the Antiquity and Diversity of Coral Endosymbionts. *Curr. Biol.* **2018**, *28*, 2570–2580. [CrossRef]

45. Weber, M.; Medina, M. The role of microalgal symbionts (*Symbiodinium*) in holobiont physiology. *Adv. Bot. Res.* **2012**, *64*, 119–140.

46. Sampayo, E.E.; Ridgway, T.; Bongaerts, P.; Hoegh-Guldberg, O. Bleaching susceptibility and mortality of corals are determined by fine-scale differences in symbiont type. *Proc. Natl. Acad. Sci. USA* **2008**, *105*, 10444–10449. [CrossRef]

47. Jones, A.M.; Berkelmans, R.; Van Oppen, M.J.H.; Mieog, J.C.; Sinclair, W. A community change in the algal endosymbionts of a scleractinian coral following a natural bleaching event: Field evidence of acclimatization. *Proc. R. Soc. B Biol. Sci.* **2008**, *275*, 1359–1365. [CrossRef] [PubMed]

48. Ulstrup, K.E.; Van Oppen, M. Geographic and habitat partitioning of genetically distinct zooxanthellae (*Symbiodinium*) in *Acropora* corals on the Great Barrier Reef. *Mol. Ecol.* **2003**, *12*, 3477–3484. [CrossRef]

49. Lien, Y.-T.; Nakano, Y.; Plathong, S.; Fukami, H.; Wang, J.-T.; Chen, C.A. Occurrence of the putatively heat-tolerant *Symbiodinium* phylotype D in high-latitudinal outlying coral communities. *Coral Reefs* **2007**, *26*, 35–44. [CrossRef]

50. Ghavam Mostafavi, P.; Fatemi, S.M.R.; Shahhosseiny, M.H.; Hoegh-Guldberg, O.; Loh, W.K.W. Predominance of clade D *Symbiodinium* in shallow-water reef-building corals off Kish and Larak Islands (Persian Gulf, Iran). *Mar. Biol.* **2007**, *153*, 25–34. [CrossRef]

51. Oliver, T.A.; Palumbi, S.R. Many corals host thermally resistant symbionts in high-temperature habitat. *Coral Reefs* **2011**, *30*, 241–250. [CrossRef]

52. Chen, C.A. Analysis of Scleractinian Distribution in Taiwan Indicating a Pattern Congruent with Sea Surface Temperatures and Currents: Examples from *Acropora* and Faviidae Corals. *Zool. Stud.* **1999**, *38*, 119–129.

53. Chen, C.A.; Keshavmurthy, S. Taiwan as a connective stepping-stone in the Kuroshio Traiangle and the conservation of coral ecosystems under the impacts of climate change. *Kuroshio Sci.* **2009**, *3*, 15–22.

54. Dai, C.F. Coastal and shallower water ecosystem. In *Chapter 7. Regional Oceanography of Taiwan Version II*; Jan, S., Ed.; National Taiwan University Press: Taipei, Taiwan, 2018; pp. 263–299.

55. Dai, C.F. Assessment of the present health of coral reefs in Taiwan. In *Status of Coral Reefs in the Pacific*; UNIHI Sea Grant CP-98-01; Grigg, R.W., Birkeland, C., Eds.; Sea Grant Program, University of Hawaii: Honolulu, HI, USA, 1997; pp. 123–131.

56. Bruno, J.F.; Selig, E.R. Regional Decline of Coral Cover in the Indo-Pacific: Timing, Extent, and Subregional Comparisons. *PLoS ONE* **2007**, *2*, e711. [CrossRef]

57. Jackson, J.B.C.; Donovan, M.K.; Cramer, K.L.; Lam, V.V. (Eds.) Status and Trends of Caribbean Coral Reefs: 1970–2012. In *Global Coral Reef Monitoring Network*; IUCN: Gland, Switzerland, 2014; 298p.

58. Meng, P.-J.; Chen, J.-P.; Chung, J.-N.; Liu, M.-C.; Fan, T.-Y.; Chang, C.-M.; Tian, W.-M.; Chang, Y.-C.; Lin, H.-J.; Fang, L.-S.; et al. Long-term ecological studies in Kenting National Park neighboring marine areas, on monitoring the impact factors from anthropogenic activities to the marine ecosystem and a preliminary database of its marine ecosystem. *J. Natl. Park* **2004**, *14*, 43–69, (In Chinese with English abstract).

59. Meng, P.-J.; Lee, H.-J.; Wang, J.-T.; Chen, C.-C.; Lin, H.-J.; Tew, K.S.; Hsieh, W.-J. A long-term survey on anthropogenic impacts to the water quality of coral reefs, southern Taiwan. *Environ. Pollut.* **2008**, *156*, 67–75. [CrossRef] [PubMed]

60. Liu, P.-J.; Meng, P.-J.; Liu, L.-L.; Wang, J.-T.; Leu, M.-Y. Impacts of human activities on coral reef ecosystems of southern Taiwan: A long-term study. *Mar. Pollut. Bull.* **2012**, *64*, 1129–1135. [CrossRef] [PubMed]

61. Chen, C.A. Status of Coral Reefs in Taiwan. In *Status of Coral Reefs in East Asian Seas Region*; Kimura, T., Tun, K., Chou, L.M., Eds.; Ministry of Environment: Tokyo, Japan, 2014; pp. 69–78.

62. Kuo, C.-Y.; Yuen, Y.-S.; Meng, P.-J.; Ho, P.-H.; Wang, J.T.; Liu, P.J.; Chang, Y.C.; Dai, C.F.; Fan, T.Y.; Lin, H.J.; et al. Recurrent disturbances and the degradation of hard coral communities in Taiwan. *PLoS ONE* **2012**, *7*, e44364. [CrossRef] [PubMed]

63. Ribas-Deulofeu, L.; Denis, V.; De Palmas, S.; Kuo, C.-Y.; Hsieh, H.J.; Chen, C.A. Structure of Benthic Communities along the Taiwan Latitudinal Gradient. *PLoS ONE* **2016**, *11*, e0160601. [CrossRef] [PubMed]

64. Lee, H.-J.; Chao, S.-Y.; Fan, K.-L.; Kuo, T.-Y. Tide-Induced Eddies and Upwelling in a Semi-enclosed Basin: Nan Wan. *Estuar. Coast. Shelf Sci.* **1999**, *49*, 775–787. [CrossRef]

65. Lee, H.-J.; Chao, S.-Y.; Fan, K.-L. Flood-ebb disparity of tidally induced recirculation Eddies in a semi-enclosed basin: Nan Wan Bay. *Cont. Shelf Res.* **1999**, *19*, 871–890. [CrossRef]

66. Sugiyama, T. Reef building corals of Japanese coast. *Pap. Inst. Geol. Paleontol. Tohoku Imp. Univ.* **1937**, *26*, 1–60.

67. Kawaguti, S. Coral fauna of the Taiwan waters. *Kagaku No Taiwan* **1942**, *11*, 1–6.

68. Kawaguti, S. Coral fauna of Garampi. *Trans. Formos. Nat. Hist. Soc.* **1943**, *33*, 258–259.

69. Kawaquti, S. Coral fauna of the island of Botel Tobago, Formosa with a list of corals from the Formosan waters. *Biol. J. Okoyama Univ.* **1953**, *1*, 185–198.

70. Ma, T.Y.H. The effect of warm and cold currents in the southern-western Pacific on the growth rate of reef corals. *Oceanogr. Sin.* **1957**, *5*, 1–34.

71. MA, T.Y.H. The relation of growth rate of reef corals to surface temperature of sea water as basis for study of causes of diastrophisms instigating evolution of life. In *Research on the Past Climate and Continental Drift*; World Book Co.: Taipei, Taiwan, 1958; Volume 14.

72. Ma, T.Y.H. Effect of water temperature on growth rate of corals. *Oceanogr. Sin. Specif.* **1959**, *1*, 116.

73. Jones, O.A.; Randall, R.H.; Cheng, Y.M.; Kami, H.T.; Mak, S.M. *A Marine Biological Survey of Southern Taiwan with Emphasis on Corals and Fishes*; Institute of Oceanography, National Taiwan University: Taipei, Taiwan, 1972; 93p.

74. Yang, R.T.; Huang, C.C.; Wang, C.H.; Yeh, S.Z.; Jan, Y.F.; Liu, S.L.; Chen, C.H.; Chen, S.J.; Chang, L.F.; Sun, C.L.; et al. *A Marine Biological Data Acquisition Program Pertaining to the Construction of a Power Plant in the Nan-Wan Bay Area. Phase, I. A Preliminary Reconnaissance Survey*; Institute of Oceanography, National Taiwan University: Taipei, Taiwan, 1976; 134p.

75. Yang, R.T.; Huang, C.C.; Wang, C.H.; Yeh, S.Z.; Jan, Y.F.; Liu, S.L.; Chen, C.H.; Chen, S.J.; Chang, L.F.; Sun, C.L.; et al. *A Marine Biological Data Acquisition Program Pertaining to the Construction of a Power Plant in the Nan-Wan Bay Area. Phasw 11, Biological Data Acquisition*; Institute of Oceanography, National Taiwan University: Taipei, Taiwan, 1977; pp. 1–34.

76. Randall, R.H.; Cheng, Y.M. Recent corals of Taiwan. Part, I. Description of reefs and coral environment. *Acta Geol. Taiwanica* **1977**, *19*, 79–102.

77. Randall, R.H.; Cheng, Y.M. Recent corals of Taiwan. Part II. Description of reefs and coral environment. *Acta Geol. Taiwanica* **1979**, *20*, 1–32.

78. Yang, R.T. Coral communities in Nan-wan Bay (Taiwan). In Proceedings of the Fifth International Coral Reef Congress, Tahiti, French Polynesia, 27 May–1 June 1985; Volume 6, pp. 273–278.

79. Dai, C.-F. Community Ecology of Corals on the Fringing Reefs of Southern Taiwan. Ph.D. Thesis, Yale University, New Haven, CT, USA, 1988; p. 315.

80. Dai, C.-F. Coral communities of southern Taiwan. In Proceedings of the 6th International Coral Reef Symposium, Townsville, Australia, 8–12 August 1988; Volume 2, pp. 647–652.

81. Dai, C.-F.; Chen, Y.-T.; Kuo, K.-M.; Chuang, C.-H. Changes of coral communities in Nanwan Bay, Kenting National Park: 1987–1997. *J. Natl. Park* **1998**, *8*, 79–99, (In Chinese with English abstract).

82. Dai, C.-F.; Kuo, K.-M.; Chen, Y.-T.; Chuang, C.-H. Changes of coral communities on the east and west coast of the Kenting National Park. *J. Natl. Park* **1999**, *9*, 112–130, (In Chinese with English abstract).

83. Chang, J.-S.; Dai, C.-F.; Chang, J. The gametangium-like structure as propagation buds in *Codium edule* Silva (Bryopsidales: Chlorophya). *Bot. Mar.* **2003**, *46*, 431–437. [CrossRef]

84. Chen, C.A.; Dai, C.-F. Local phase shift from Acropora-dominant to *Condylactis-* dominant community in the Tiao-Shi Reef, Kenting National Park, southern Taiwan. *Coral Reefs* **2004**, *23*, 508. [CrossRef]

85. Shao, K.-T.; Fang, L.-S.; Liang, N.-K.; Meng, P.-J.; Chung, K.-N.; Li, J.-J.; Han, C.-C. *Long-Term Ecological Research (LTER) Program of the Coral Reefs in Kenting National Park, Taiwan*; Kenting National Park Headquarters, Construction and Planning Agency, Minister of the Interior: Taipei, Taiwan, 2002; p. 176. (In Chinese)

86. Fang, L.-S.; Shao, K.-T.; Tian, W.-M.; Chang, Y.-C.; Lin, H.-J.; Fan, T.-Y.; Chung, K.-N.; Chen, J.-P.; Liu, M.-Y.; Meng, P.-J. *Long-Term Ecological Research (LTER) Program of the Coral Reefs in Kenting National Park, Taiwan*; Kenting National Park Headquarters, Construction and Planning Agency, Minister of the Interior: Taipei, Taiwan, 2003; p. 249. (In Chinese)

87. Fang, L.-S.; Shao, K.-T.; Tian, W.-M.; Chang, Y.-C.; Lin, H.-J.; Fan, T.-Y.; Chung, K.-N.; Chen, J.-P.; Chen, M.-H.; Liu, M.-Y.; et al. *Long-Term Ecological Research (LTER) Program of the Coral Reefs in Kenting National Park, Taiwan*; Kenting National Park Headquarters, Construction and Planning Agency, Minister of the Interior: Taipei, Taiwan, 2004; p. 262. (In Chinese)

88. Fang, L.-S.; Shao, K.-T.; Meng, P.-J.; Chung, K.-N.; Chen, J.-P.; Chen, M.-H.; Liu, M.-Y.; Chang, Y.-C.; Fan, T.-Y.; Lin, H.-J. *Long-Term Ecological Research (LTER) Program of the Coral Reefs in the Kenting National Park*; Kenting National Park Headquarters, Construction and Planning Agency, Minister of the Interior: Taipei, Taiwan, 2005; p. 295. (In Chinese)

89. Fang, L.-S.; Shao, K.-T.; Meng, P.-J.; Chung, K.-N.; Chen, J.-P.; Chen, M.-H.; Liu, M.-Y.; Chang, Y.-C.; Fan, T.-Y.; Lin, H.-J. *Long-Term Ecological Research (LTER) Program of the Coral Reefs in Kenting National Park, Taiwan*; Kenting National Park Headquarters, Construction and Planning Agency, Minister of the Interior: Taipei, Taiwan, 2006; p. 233. (In Chinese)

90. Wang, W.-H.; Shao, K.-T.; Meng, P.-J.; Fan, T.-Y.; Chen, J.-P.; Chen, M.-H.; Liu, M.-Y.; Chang, Y.-C.; Lin, H.-J.; Ho, P.-H. *Long-Term Ecological Research (LTER) Program of the Coral Reefs in Kenting National Park, Taiwan*; Kenting National Park Headquarters, Construction and Planning Agency, Minister of the Interior: Taipei, Taiwan, 2007; p. 226. (In Chinese)

91. Ho, M.-J.; Cheng, N.-Y.; Chen, Y.-C.; Kuo, C.-Y.; Wen, C.K.-C.; Cherh, K.-L.; Chen, C.A. Investigating current status of benthic ecology in the Kenting National Park marine zones, Taiwan. *J. Natl. Park* **2016**, *26*, 46–56, (In Chinese with English abstract).

92. Ho, P.-H.; Chen, C.A.; Chen, H.-Y.; Chen, J.-P.; Chiu, Y.-W.; Ho, P.-H.; Lin, H.-J.; Chang, Y.-C. *Long-Term Ecological Research (LTER) Program of the Coral Reefs in Kenting National Park, Taiwan*; Kenting National Park Headquarters, Construction and Planning Agency, Minister of the Interior: Taipei, Taiwan, 2008; p. 270. (In Chinese)

93. Ho, P.-H.; Chen, C.A.; Meng, P.-J.; Chen, J.-P.; Chiu, Y.-W.; Lin, H.-J.; Chang, Y.-C. *Long-Term Ecological Research (LTER) Program of the Coral Reefs in Kenting National Park, Taiwan*; Kenting National Park Headquarters, Construction and Planning Agency, Minister of the Interior: Taipei, Taiwan, 2009; p. 276. (In Chinese)

94. Ho, P.-H.; Chen, C.A.; Meng, P.-J.; Chen, J.-P.; Chiu, Y.-W.; Lin, H.-J.; Chang, Y.-C.; Liu, P.-J. *Long-Term Ecological Research (LTER) Program of the Coral Reefs in Kenting National Park, Taiwan*; Kenting National Park Headquarters, Construction and Planning Agency, Minister of the Interior: Taipei, Taiwan, 2010; p. 300. (In Chinese)

95. Ho, P.-H.; Chen, C.A.; Meng, P.-J.; Chen, J.-P.; Chiu, Y.-W.; Lin, H.-J.; Chang, Y.-C.; Liu, P.-J. *Long-Term Ecological Research (LTER) Program of the Coral Reefs in Kenting National Park, Taiwan*; Kenting National Park Headquarters, Construction and Planning Agency, Minister of the Interior: Taipei, Taiwan, 2011; p. 330. (In Chinese)

96. Chen, C.A. *Long-Term Ecological Research (LTER) Program of the Coral Reefs in Kenting National Park, Taiwan*; Kenting National Park Headquarters, Construction and Planning Agency, Minister of the Interior: Taipei, Taiwan, 2012; p. 73. (In Chinese)

97. Chen, C.A.; Meng, P.-J.; Chiu, Y.-W. *Annual Monitoring of Benthic Community, Water Quality and Shellfish in Kenting National Park*; Kenting National Park Headquarters, Construction and Planning Agency, Minister of the Interior: Taipei, Taiwan, 2013; p. 119. (In Chinese)

98. Chen, C.A. *Evaluating and Re-Design the Zoning System in Kenting National Park, Taiwan*; Kenting National Park Headquarters, Construction and Planning Agency, Minister of the Interior: Taipei, Taiwan, 2014; p. 133. (In Chinese)

99. Chen, C.A. *Annual Survey of the Biodiversity of Coral Reef Ecosystem in Kenting National Park, Taiwan*; Kenting National Park Headquarters, Construction and Planning Agency, Minister of the Interior: Taipei, Taiwan, 2016; p. 82. (In Chinese)

100. Chen, C.A. *Annual Survey of the Biodiversity of Coral Reef Ecosystem in Kenting National Park, Taiwan*; Kenting National Park Headquarters, Construction and Planning Agency, Minister of the Interior: Taipei, Taiwan, 2018; p. 100. (In Chinese)

101. Kuo, C.-Y. The Structure and Variation of Benthic Communities in Coral Reefs of Southern Taiwan. Master's Thesis, National Sun Yat-sen University, Kaohsiung, Taiwan, 2007; p. 71, (In Chinese with English abstract).

102. Kuo, C.-Y.; Meng, P.-J.; Ho, P.-H.; Wang, J.-T.; Chen, J.-P.; Chiu, Y.-W.; Lin, H.-J.; Chang, Y.-C.; Fan, T.-Y.; Chen, C.M. Damage to the reefs of Siangjiao Bay marine protected area in Kenting National Park, Taiwan during Typhoon Morakot. *Zool. Stud.* **2011**, *50*, 85.

103. Kuo, C.-Y.; Ho, M.-J.; Chen, Y.-C.; Yang, S.-Y.; Huang, Y.-Y.; Hsieh, H.-J.; Jeng, M.-S.; Chen, C.A. *Status of Coral Reefs in East Asian Seas Region: Taiwan*; Ministry of the Environment of Japan; Japan Wildlife Research Center: Tokyo, Japan, 2018; pp. 33–40.

104. Kohler, K.E.; Gill, S.M. Coral point count with Excel extensions (CPCe): A visual basic program for the determination of coral and substrate coverage using random point count methodology. *Comput. Geosci.* **2006**, *32*, 1259–1269. [CrossRef]

105. Price, J.F. Upper ocean response to a hurricane. *J. Phys. Oceanogr.* **1981**, *11*, 153–175. [CrossRef]

106. Wada, A.; Uehara, T.; Ishizaki, S. Typhoon-induced sea surface cooling during the 2011 and 2012 typhoon seasons: Observational evidence and numerical investigations of the sea surface cooling effect using typhoon simulations. *Prog. Earth Planet. Sci.* **2014**, *64*, 3562–3578. [CrossRef]

107. Keshavmurthy, S.; Meng, P.-J.; Wang, J.-T.; Kuo, C.-Y.; Yang, S.-Y.; Hsu, C.-M.; Gan, C.-H.; Dai, C.-F.; Chen, C.A. Can resistant coral-*Symbiodinium* associations enable coral communities to survive climate change? A study of a site exposed to long-term hot water input. *PeerJ* **2014**, *2*, e327. [CrossRef] [PubMed]

108. Hsu, C.-M.; Keshavmurthy, S.; Denis, V.; Kuo, C.-Y.; Wang, J.-T.; Meng, P.-J.; Chen, C.A. Temporal and Spatial Variations in Symbiont Communities of Catch Bowl Coral *Isopora palifera* (Scleractinia: Acroporidae) on Reefs in Kenting National Park, Taiwan. *Zool. Stud.* **2012**, *51*, 1343–1353.

109. Keshavmurthy, S.; Hsu, C.-M.; Kuo, C.-Y.; Meng, P.-J.; Wang, J.-T.; Chen, C.A. Symbiont communities and host genetic structure of the brain coral *Platygyra verweyi*, at the outlet of a nuclear power plant and adjacent areas. *Mol. Ecol.* **2012**, *21*, 4393–4407. [CrossRef]

110. Carballo-Bolaños, R.; Denis, V.; Huang, Y.-Y.; Keshavmurthy, S.; Chen, C.A. Temporal variation and photochemical efficiency of species in Symbiodinaceae associated with coral *Leptoria phrygia* (Scleractinia; Merulinidae) exposed to contrasting temperature regimes. *PLoS ONE* **2019**, *14*, e0218801. [CrossRef]

111. Chiou, W.-D.; Cheng, L.-Z.; Ou, H.-C. Relationship between the dispersion of thermal effluent and the tidal current in the waters near the outlet of the third nuclear power plant in southern Taiwan. *J. Fish. Soc. Taiwan* **1993**, *20*, 207–220.

112. Fan, K.-L. The thermal effluent problems of three nuclear power plants in Taiwan. In *Oceanography of Asian Marginal Seas*, *54*; Takano, K., Ed.; Elsevier: Amsterdam, The Netherlands, 1991; pp. 393–403.

113. Pier, J.-J. Power uprate effect on thermal effluent of nuclear power plants in Taiwan. In *Nuclear Power—Operation, Safety and Environment*; Tsvetkov, P., Ed.; InTech Publication: Rijeka, Croatia, 2011; 287p.

114. Chen, C.A.; Yang, Y.W.; Wei, N.V.; Tsai, W.S.; Fang, L.S. Symbiont diversity in scleractinian corals from tropical reefs and subtropical non-reef communities in Taiwan. *Coral Reefs* **2005**, *24*, 11–22. [CrossRef]

115. Huang, Y.-Y.; Carballo-Bolaños, R.; Kuo, C.-Y.; Keshavmurthy, S.; Chen, C.A. Seasonal dynamics of the Symbiodiniaceae in Leptoria phrygia (Scleractinia; Merulinidae). (submitted, under review).

116. Fong, W.-L. Spatial and Temporal Variations of Host Genetic Structure and Symbiont Community in the Catch Bowl Coral, *Isopora Palifera* (Scleractinia; Acroporidae), in Kenting National Park, Taiwan. Master's Thesis, Institute of Oceanography, National Taiwan University, Taipei, Taiwan, 2016.

117. Baker, A.C.; Starger, C.J.; McClanahan, T.R.; Glynn, P.W. Coral reefs: Corals' adaptive response to climate change. *Nature* **2004**, *430*, 741. [CrossRef]

118. Berkelmans, R.; Van Oppen, M. The role of zooxanthellae in the thermal tolerance of corals: A 'nugget of hope' for coral reefs in an era of climate change. *Proc. R. Soc. B Biol. Sci.* **2006**, *273*, 2305–2312. [CrossRef] [PubMed]

119. Silverstein, R.N.; Cunning, R.; Baker, A.C. Change in algal symbiont communities after bleaching, not prior heat exposure, increases heat tolerance of reef corals. *Glob. Chang. Biol.* **2015**, *21*, 236–249. [CrossRef] [PubMed]

120. Cunning, R.; Silverstein, R.N.; Baker, A.C. Investigating the causes and consequences of symbiont shuffling in a multi-partner reef coral symbiosis under environmental change. *Proc. R. Soc. B Biol. Sci.* **2015**, *282*, 20141725. [CrossRef]

121. Boulotte, N.M.; Dalton, S.J.; Carroll, A.G.; Harrison, P.L.; Putnam, H.M.; Peplow, L.M.; van Oppen, M.J. Exploring the *Symbiodinium* rare biosphere provides evidence for symbiont switching in reef-building corals. *ISME J.* **2016**. [CrossRef] [PubMed]

122. Thornhill, D.J.; Lajeunesse, T.C.; Kemp, D.W.; Fitt, W.K.; Schmidt, G.W. Multi-year, seasonal genotypic surveys of coral-algal symbioses reveal prevalent stability or post-bleaching reversion. *Mar. Biol.* **2006**, *148*, 711–722. [CrossRef]

123. Keshavmurthy, S.; Tang, K.-H.; Hsu, C.-M.; Gan, C.-H.; Kuo, C.-Y.; Soong, K.; Chou, H.-N.; Chen, C.A. *Symbiodinium* spp. associated with scleractinian corals from Dongsha Atoll (Pratas), Taiwan, in the South China Sea. *PeerJ* **2017**, *5*, e2871. [CrossRef]

124. Tong, H.; Cai, L.; Zhou, G.; Yuan, T.; Zhang, W.; Tian, R.; Huang, H.; Qian, P.-Y. Temperature shapes coral-algal symbiosis in the South China Sea. *Sci. Rep.* **2017**, *7*, 40118. [CrossRef] [PubMed]

125. Gong, S.; Chai, G.; Xiao, Y.; Xu, L.; Yu, K.; Li, J.; Liu, F.; Cheng, H.; Zhang, F.; Liao, B.; et al. Flexible symbiotic associations of *Symbiodinium* with five typical coral species in tropical and subtropical reef regions of the northern South China Sea. *Front. Microbiol.* **2018**, *9*, 2485. [CrossRef] [PubMed]

126. Chen, C.A.; Wang, J.-T.; Fang, L.-S.; Yang, Y.-W. Fluctuating algal symbiont communities in *Acropora palifera* (Scleractinia: Acroporidae) from Taiwan. *Mar. Ecol. Prog. Ser.* **2005**, *295*, 113–121. [CrossRef]

127. Thornhill, D.J.; Fitt, W.K.; Schmidt, G.W. Highly stable symbioses among western Atlantic brooding corals. *Coral Reefs* **2006**, *25*, 515–519. [CrossRef]

128. Tonk, L.; Sampayo, E.E.; Lajeunesse, T.C.; Schrameyer, V.; Hoegh-Guldberg, O. *Symbiodinium* (Dinophyceae) diversity in reef-invertebrates along an offshore to inshore reef gradient near Lizard Island, Great Barrier Reef. *J. Phycol.* **2014**, *50*, 552–563. [CrossRef] [PubMed]

129. Beyer, H.L.; Kenndey, E.V.; Beger, M.; Chen, C.A.; Cinner, J.E.; Darling, E.S.; Eakin, C.M.; Gates, R.D.; Heron, S.F.; Knowlton, N.; et al. Risk-sensitive planning for conserving coral reefs under rapid climate change. *Conserv. Lett.* **2018**, *11*, e12587. [CrossRef]

130. Ablan, M.C.A.; McManus, J.W.; Chen, C.A.; Shao, K.T.; Bell, J.; Cabanban, A.S.; Tuan, V.S.; Arthana, I.W. Meso-scale transboundary units for the management of coral reefs in the South China Sea area. *NAGA Worldfish Cent. Q.* **2002**, *25*, 4–9.

131. Chen, C.A.; Ablan, M.C.A.; McManus, J.W.; Bell, J.D.; Tuan, V.S.; Cabanban, A.S.; Shao, K.-T. Population structure and genetic variability of six-bar wrasse (*Thallasoma hardwicki*) in northern South China Sea revealed by mitochondrial control region sequences. *Mar. Biotechnol.* **2004**, *6*, 312–326. [CrossRef]

132. Chou, L.M. Southeast Asian reefs-a status update: Cambodia, Indonesia, Malaysia, Philippines, Singapore, Thailand and Vietnam. In *Status of the Coral Reefs of the World: 2000*; Wilkinson, C., Ed.; Australian Institute of Marine Science: Townsville, Australia, 2000; pp. 117–129.

133. West, J.M.; Slam, R.V. Resistance and resilience to coral bleaching: Implications for coral reef conservation and management. *Conserv. Biol.* **2003**, *4*, 956–967. [CrossRef]

134. Riegl, B.; Piller, W.E. Possible refugia for reefs in times of environmental stress. *Int. J. Earth Sci.* **2003**, *92*, 520. [CrossRef]

135. Grimsditch, G.D.; Salm, R.V. *Coral Reef Resilience and Resistance to Bleaching*; IUCN: Gland, Switzerland, 2005; 50p.

136. Chollett, I.; Mumby, P.J.; Cortes, J. Upwelling areas to not guarantee refuge for corals in a warming ocean. *Mar. Ecol. Prog. Ser.* **2010**, *416*, 47–56. [CrossRef]

137. Glynn, P.W.; D'croz, L. Experimental evidence for high temperature stress as the cause of El Nino-coincident coral mortality. *Coral Reefs* **1990**, *8*, 181–191. [CrossRef]

J. Mar. Sci. Eng. **2019**, *7*, 388

138. Sánchez-Noguera, C.; Stuhldreier, I.; Cortés, J.; Jiménez, C.; Morales, Á.; Wild, C.; Rixen, T. Natural ocean acidification at Papagayo upwelling system (North Pacific Costa Rica): Implications for reef development. *Biogeosciences* **2018**, *15*, 2349–2360. [CrossRef]
139. Wizemann, A.; Nandini, S.D.; Stuhldreier, I.; Sánchez-Noguera, C.; Wisshak, M.; Westphal, H.; Rixen, T.; Wild, C.; Reymond, C.E. Rapid bioerosion in a tropical upwelling coral reef. *PLoS ONE* **2018**, *13*, e0202887. [CrossRef]

Journal of
*Marine Science
and Engineering*

Correction: Keshavmurthy, S., et al. Coral Reef Resilience in Taiwan: Lessons from Long-Term Ecological Research on the Coral Reefs of Kenting National Park (Taiwan). *Journal of Marine Science and Engineering* 2019, 7, 338

Shashank Keshavmurthy [1,†]**, Chao-Yang Kuo** [1,†]**, Ya-Yi Huang** [1]**, Rodrigo Carballo-Bolaños** [1,2,3]**, Pei-Jei Meng** [4,5,*]**, Jih-Terng Wang** [6,*] **and Chaolun Allen Chen** [1,2,3,7,8,*]

[1] Biodiversity Research Center, Academia Sinica, Taipei 11529, Taiwan; coralresearchtaiwan@gmail.com (S.K.); cykuo.tw@gmail.com (C.-Y.K.); lecy.yhuang@gmail.com (Y.-Y.H.); rodragoncar@gmail.com (R.C.-B.)
[2] Biodiversity Program, Taiwan International Graduate Program, Academia Sinica, Taipei 11529, Taiwan
[3] Department of Life Science, National Taiwan Normal University, Taipei 10610, Taiwan
[4] National Museum of Marine Biology/Aquarium, Pintung 94450, Taiwan
[5] Insititute of Marine Biodiversity and Evolution, National Dong-Hua University, Hualien 97447, Taiwan
[6] Department of Biotechnology, Tajen University, Pingtung 90741, Taiwan
[7] Institute of Oceanography, National Taiwan University, Taipei 10617, Taiwan
[8] Department of Life Science, Tunghai University, Taichung 40704, Taiwan
[*] Correspondence: pjmeng@nmmba.gov.tw (P.-J.M.); jtwtaiwan@gmail.com (J.-T.W.); cac@gate.sinica.edu.tw (C.A.C.)
[†] These two authors contributed equally to the work described.

Received: 4 August 2020; Accepted: 15 September 2020; Published: 26 November 2020

The authors are sorry for errors in their paper [1], which will not affect the interpretation or final results, but will lead to confusion. Consequently the authors wish to make the following corrections to the paper:

Change in Main Body Paragraphs

➤ "Sianjiaowan" and "Siangjiao Bay" to be replaced with "Siangjiaowan". "Jialuoshui" replaced with "Jialeshuei", "Lidashih" replaced with "Leidashih", "Longken" replaced with "Longkeng", "Dingnaisha" replaced with "Dingbaisha", SIW replaced with SJW; exchange the citation of "Figure 2" with "Figure 3", "Figure 5c" replaced with "Figure 5e", Figure 6" replaced with "Figure 7", replace "Figures 7 and 8" with "Figures 8 and 9", replace "Figure 7" with "Figure 8".
➤ Change the citation of "Table 2" to "Table 3", Table 3" to "Table 4, "Table 4" to "Table 2".
➤ Insert "in Taioshi" between [84] and Condylactis sp.

Change in Figures/Tables

➤ Replace Figure 2:

With new Figure 2 below:

➢ Due to mislabeling, replace Figure 5:

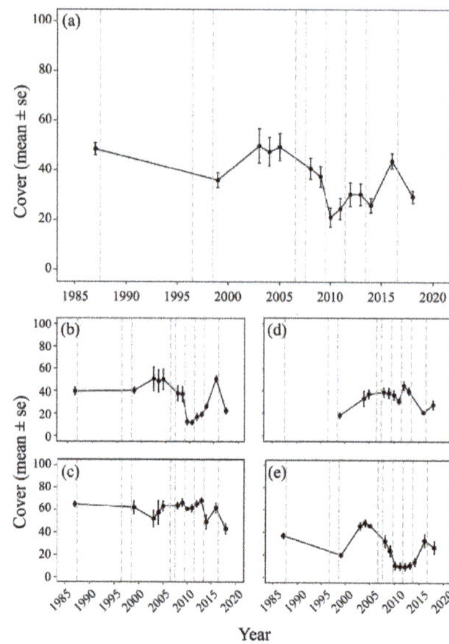

With new Figure 5:

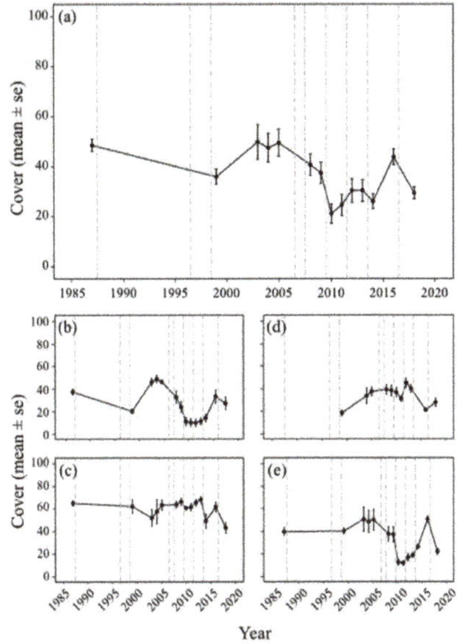

➤ Due to the wrong pie chart, replace Figure 9:

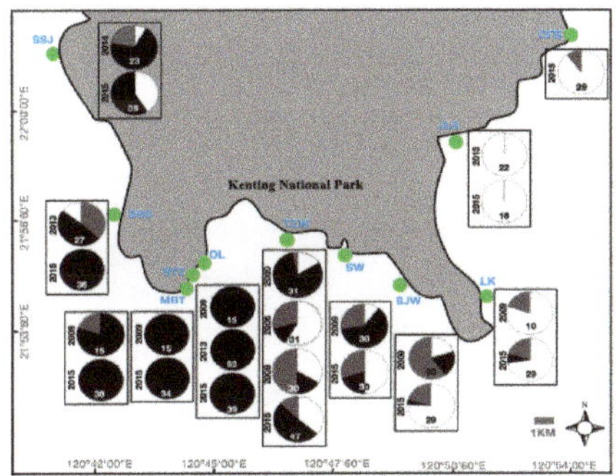

With new Figure 9 below:

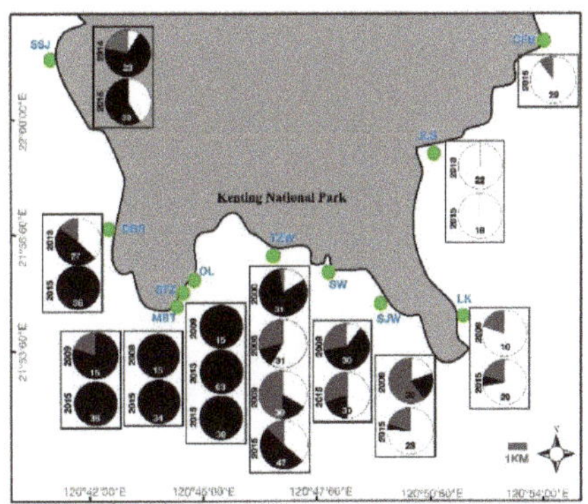

J. Mar. Sci. Eng. **2020**, *8*, 964

Reference

1. Keshavmurthy, S.; Kuo, C.-Y.; Huang, Y.-Y.; Carballo-Bolaños, R.; Meng, P.-J.; Wang, J.-T.; Chen, C.A. Coral Reef Resilience in Taiwan: Lessons from Long-Term Ecological Research on the Coral Reefs of Kenting National Park (Taiwan). *J. Mar. Sci. Eng.* **2019**, *7*, 388. [CrossRef]

Publisher's Note: MDPI stays neutral with regard to jurisdictional claims in published maps and institutional affiliations.

MDPI
St. Alban-Anlage 66
4052 Basel
Switzerland
Tel. +41 61 683 77 34
Fax +41 61 302 89 18
www.mdpi.com

Journal of Marine Science and Engineering Editorial Office
E-mail: jmse@mdpi.com
www.mdpi.com/journal/jmse